Cambridge
Collections

W0007709

The living planet

a collection of writing on the environment

Edited by Mary Green
Series editor: Michael Marland

CAMBRIDGE
UNIVERSITY PRESS

In memory of Michael Marland (1934–2008).

CAMBRIDGE UNIVERSITY PRESS
Cambridge, New York, Melbourne, Madrid, Cape Town, Singapore, São Paulo, Delhi

Cambridge University Press
The Edinburgh Building, Cambridge CB2 8RU, UK

www.cambridge.org
Information on this title: www.cambridge.org/9780521747646

© Cambridge University Press 2009

This publication is in copyright. Subject to statutory exception and to the provisions of relevant collective licensing agreements, no reproduction of any part may take place without the written permission of Cambridge University Press.

First published 2009

Printed in the United Kingdom at the University Press, Cambridge

A catalogue record for this publication is available from the British Library

ISBN 978-0-521-74764-6 paperback

Cover image: Stocktrek Images, Inc. / Alamy
Cover design by Smith
Illustrations by Julia Pearson / Beehive Illustration

Cambridge University Press has no responsibility for the persistence or accuracy of URLs for external or third-party Internet websites referred to in this publication, and does not guarantee that any content on such websites is, or will remain, accurate or appropriate. Information regarding prices, travel timetables and other factual information given in this work are correct at the time of first printing but Cambridge University Press does not guarantee the accuracy of such information thereafter.

Contents

General introduction

What does the title *The living planet* conjure up in your mind? Do you think of that well-known satellite image of the beauty of Earth from space? An explorer might think of the immensity of the natural world and an environmentalist of how to take care of it. This collection looks at life on the planet from many different aspects and contains a broad range of texts. Inside its covers you'll find examples of stories, poems, reportage, letters, drama, and passages from novels, autobiographies and non-fiction texts. It also includes an example of graphic fiction.

The collection is meant to be read for enjoyment, but also to stimulate thought, sometimes about complex, challenging issues that I hope will lead to discussion and argument and an exchange of opinion. There are useful footnotes on some pages explaining difficult words, and at the end of each section are recommendations for further reading. There are also accompanying activities that encourage close reading of the text, so that you can explore how writers and artists shape and present their feelings and ideas. Through this exploration and other supporting activities you will be able to create your own work.

The living planet is divided into four sections. In each section the easier texts are at the beginning and the more difficult towards the end. To support your reading, certain words (these are numbered) in the texts are explained in the footnotes. Ideas for further reading accompany each text, and notes about the authors can be found at the end of the anthology. Each section concludes with a range of reading, writing, speaking, listening and drama activities to help you explore and enjoy the authors' ideas, opinions, style, language and techniques.

The text-specific activities pages are divided into the following activity types: *Before you read* (pre-reading stimulation activities), *What's it about?* (comprehension-style questions) and *Thinking about the text* (activities which move beyond the text itself). At the very end of each section, a series of *Compare and contrast* activities provide opportunities to compare two or more texts.

The first of the four sections, *Wild things*, is about the abundance of wildlife on Earth. The opening text is an account of our near relative, the chimpanzee, as seen through the naturalist Jane Goodall's eyes. She has spent a lifetime studying them. Smaller mammals, reptiles, birds and fish can be found in stories and poems throughout this section and there is an extract from the play *Kes*, which itself is adapted from Barry Hines' famous book *A Kestrel for a Knave*. Underpinning much of this writing is our relationship with living creatures; how we treat them, whether or not we respect them and the effect they have on us.

Water worlds looks at water in the natural world and the way in which we are drawn to it. When you think of water you may see the great rivers, the sea and its tides, or the unknown depths. In this section you can read poems that mirror water's compelling sounds and rhythms and an extract from *The Highest Tide*, the focus of which is the mysteries of the mudflats of Puget Sound. Other texts, such as *Gone to Sea*, explore water's power to move and transform us.

The third section, *Nature's power*, is about the dramatic force of nature, and how it can strike us off-guard. There is a poem by Karen Hesse that recreates the experience of a violent dust storm, and poems by John Clare and Ted Hughes that explore the power of the wind in the English countryside. In a striking story by Jack London, the critical ingredient is the biting cold, 'seventy-five below zero'. The extreme force of nature is not always as evident in temperate climates as it is in other parts of the world, although this is already altering, as weather patterns change and we experience violent events more often.

This leads us into the final section, *Future planet*. Here the focus shifts from texts written in celebration and awe of nature to texts that are concerned about the state of the planet. In their different ways they explore what the future could be for Earth, its creatures and humanity in the face of climate change. *Green Boy* and *Floodland* take an imaginative leap into the future, and you can speculate about how the characters will overcome the challenges posed. Questions about what climate change really means are answered in *The Rough Guide to Climate Change*. The loss of a species, the Spix's Macaw, is explored in Simon Armitage's beautiful poem *The Final Straw*. The same macaw also figures in *How Many Light Bulbs Does it Take to Change a Planet?*, in which Tony Juniper outlines what we can do to help lessen the effects of a changing climate. This last point takes us to the heart of the matter, for change needs to occur at a global level, and in the way we lead our lives. In this section you can read suggestions for how both can be achieved.

The collection as a whole is intended to encourage a love of the natural world and a realisation that we are much closer to it than we often think, even in our great cities. Whether we know it or not, it has a profound bearing on our lives. We need therefore to care for it, for its own sake and for future generations. Some of you will already be doing so.

Mary Green

1 Wild things

The texts in this section explore our relationships with wildlife, through fiction, non-fiction, poetry and drama. Some, such as David Attenborough's *Life in Cold Blood*, reveal a lifelong commitment to the study of animals and their capacity for adaptation and survival. Others show humans pitted against creatures or concerned for their welfare. Most texts are celebratory, and all reveal a fascination with the untamed.

As you read these texts, you can consider:
- the diversity of living creatures
- our close connection with wildlife
- our responsibility to the natural world.

Activities

1 Wildlife television programmes have remained popular since television was first broadcast. Draw up a list of reasons why you think so many people watch them.

2 **a** People say that domestic animals are only partly tamed, that they soon resort to the wild if they need to. What wild features can you recognise in your own pets or your friends' pets?

 b The Animal Welfare Act was passed in England in 2006 and emphasises that both domestic and wild animals have needs and entitlements. In what ways do you think animals are entitled to the same treatment as humans?

3 We often give animals human traits. For example, we think of pandas as cuddly. Work in a small group and think of four animals to which we give human traits. Now consider why this is unfair to the animal and how it might affect our treatment of them.

4 It is often said that humans are the most dangerous of all animals. Conduct a class debate in which you discuss whether this is true, and why.

My Life with the Chimpanzees

by Jane Goodall

Jane Goodall has spent most of her life in Gombe National Park, in Africa, studying colonies of chimpanzees. In this extract from her autobiography she recounts how she first made contact with the chimps – or rather, how the chimps made contact with her.

Soon after she'd left, I got back one evening and was greeted by an excited Dominic. He told me that a big male chimp had spent an hour feeding on the fruit of one of the oil-nut palms growing in the camp clearing. Afterward he had climbed down, gone over to my tent, and taken the bananas that had just been put there for my supper.

This was fantastic news. For months the chimps had been running off when they saw me – now one had actually visited my camp! Perhaps he would come again.

The next day I waited, in case he did. What a luxury to lie in until 7:00 A.M. As the hours went by I began to fear that the chimp wouldn't come. But finally, at about four in the afternoon, I heard a rustling in the undergrowth opposite my tent, and a black shape appeared on the other side of the clearing.

I recognized him at once. It was the handsome male with the dense white beard. I had already named him David Greybeard. Quite calmly he climbed into the palm and feasted on its nuts. And then he helped himself to the bananas I had set out for him.

There were ripe palm nuts on that tree for another five days, and David Greybeard visited three more times and got lots of bananas.

A month later, when another palm tree in camp bore ripe fruit, David again visited us. And on one of those occasions he actually took a banana from my hand. I could hardly believe it.

From that time on things got easier for me. Sometimes when I met David Greybeard out in the forest, he would come up to see if I had a banana hidden in my pocket. The other chimps stared with amazement. Obviously I wasn't as dangerous as they had thought. Gradually they allowed me closer and closer.

It was David Greybeard who provided me with my most exciting observation. One morning, near the Peak, I came upon him squatting on a termite mound. As I watched, he picked a blade of grass, poked it into a tunnel in the mound, and then withdrew it. The grass was covered with termites all clinging on with their jaws. He picked them off with his lips and scrunched them up. Then he fished for more. When his piece of grass got bent, he dropped it, picked up a little twig, stripped the leaves off it, and used that.

I was really thrilled. David had used objects as tools! He had also changed a twig into something more suitable for fishing termites. He had actually *made* a tool. Before this observation, scientists had thought that only humans could make tools. Later I would learn that chimpanzees use more objects as tools

than any creature except for us. This finding excited Louis Leakey more than any other.

In October the dry season ended and it began to rain. Soon the golden mountain slopes were covered with lush green grass. Flowers appeared, and the air smelled lovely. Most days it rained just a little. Sometimes there was a downpour. I loved being out in the forest in the rain. And I loved the cool evenings when I could lace the tent shut and make it cozy inside with a storm lantern. The only trouble was that everything got damp and grew mold. Scorpions and giant poisonous centipedes sometimes appeared in the tent – even, a few times, a snake. But I was lucky – I never got stung or bitten.

The chimpanzees often seemed miserable in the rain. They looked cold, and they shivered. Since they were clever enough to use tools, I was surprised that they had not learned to make shelters. Many of them got coughs and colds. Often, during heavy rain, they seemed irritable and bad tempered.

Once, as I walked through thick forest in a downpour, I suddenly saw a chimp hunched in front of me. Quickly I stopped. Then I heard a sound from above. I looked up and there was a big chimp there, too. When he saw me he gave a loud, clear wailing *wraaaah* – a spine-chilling call that is used to threaten a dangerous animal. To my right I saw a large black hand shaking a branch and bright eyes glaring threateningly through the foliage. Then came another savage *wraaaah* from behind. Up above, the big male began to sway the vegetation. I was surrounded. I crouched down, trying to appear as non-threatening as possible.

Suddenly a chimp charged straight toward me. His hair bristled with rage. At the last minute he swerved and ran off. I stayed still. Two more chimps charged nearby. Then, suddenly, I realized I was alone again. All the chimps had gone.

Only then did I realize how frightened I had been. When I stood up my legs were trembling! Male chimps, although they are only four feet tall when upright, are at least three times stronger than a grown man. And I weighed only about ninety pounds. I had become very thin with so much climbing in the mountains and only one meal a day. That incident took place soon after the chimps had lost their initial terror of me but before they had learned to accept me calmly as part of their forest world. If David Greybeard had been among them, they probably would not have behaved like that, I thought.

After my long days in the forests I looked forward to supper. Dominic always had it ready for me when I got back in the evenings. Once a month he went into Kigoma[1] with Hassan. They came back with new supplies, including fresh vegetables and fruit and eggs. And they brought my mail – that was something I really looked forward to.

After supper I would get out the little notebook in which I had scribbled everything I had seen while watching the chimps during the day. I would settle down to write it all legibly into my journal. It was very important to do that every evening, while it was all fresh in my mind. Even on days when I climbed back to sleep near the chimps, I always wrote up my journal first.

Gradually, as the weeks went by, I began to recognize more and more chimpanzees as individuals. Some, like Goliath, William, and old Flo, I got to know well, because David Greybeard sometimes brought them with him when he visited camp. I always had a supply of bananas ready in case the chimps arrived.

Once you have been close to chimps for a while they are as easy to tell apart as your classmates. Their faces look different, and they have different characters. David Greybeard, for example, was a calm chimp who liked to keep out of trouble. But he was also very determined to get his own way. If he arrived in camp and couldn't find any bananas, he would walk into my

---

[1]**Kigoma**   a town and port (by a lake) in Western Tanzania

tent and search. Afterward, all was chaos. It looked as though some burglar had raided the place! Goliath had a much more excitable, impetuous temperament. William, with his long-shaped face, was shy and timid.

Old Flo was easy to identify. She had a bulbous nose and ragged ears. She came to camp with her infant daughter, whom I named Fifi, and her juvenile son, Figan. Sometimes adolescent Faben came, too. It was from Flo that I first learned that in the wild, female chimps have only one baby every five or six years. The older offspring, even after they have become independent, still spend a lot of time with their mothers, and all the different family members help one another.

Further reading

Have you seen the film *Gorillas in the Mist*, in which Sigourney Weaver plays Dian Fossey, a naturalist who worked with mountain gorillas? The film was adapted from the book of the same name written by Dian Fossey (Phoenix, 2001), which you may like to read. Goodall is also the author of *Africa in My Blood: an Autobiography in Letters: The Early Years* (Houghton Mifflin, 2001) about the start of her life in Africa.

Maninagar Days

by Sujata Bhatt

Several of Sujata Bhatt's poems bridge the gap between humans and wildlife. The verses below come from the longer poem *Maninagar Days*. In it she depicts the emotional impact that wildlife can have in childhood.

They are always there
just as pigeons or flies
can be *always there*
and the children have to fight them off,
especially during those hot May afternoons
when they dare to jump down from the trees
into the cool shaded spots, the corners between
the canna flower beds
still moist from the mornings watering.

Monkeys in the garden –
I'm talking about rhesus monkeys[1]
the colour of dirt roads and khaki
         and sometimes even of honey.
Rhesus monkeys that travel in small groups,
extended families; constantly feuding brothers, sisters,
uncles, aunts, cousins screaming through
the trees – while the grandmother sits farther away
sadly, holding on to the sleepy newborn.
Somehow they manage to make peace
Before every meal.

---

[1]**Rhesus monkey**   a small brown or grey monkey, with a pink face

Now and then a solitary langur;
the Hanuman-monkey,[2] crossing the terrace
with the importance of someone going to the airport.
A lanky dancer's steps
with black hands, black feet
sharp as black leather gloves and black leather shoes
against the soft grey body.
Sharp
and yet delicate
as if they were brushed-stroked in
with a Japanese flourish.
And black-faced too,
with thick tufts of silver grey eyebrows,
a bushy chin. So aloof
he couldn't be bothered
with anyone.

Further reading

Try reading the complete poem *Maninagar Days*, in which the relation-ship between the monkeys and the children is explored more deeply. You will find it in *Monkey Shadows* (Carcanet Press, 1991).

What other poems do you know about animals? You could read some in the *Faber Book of Beasts*, compiled by Paul Muldoon (Faber and Faber, 1998). There are numerous examples.

---

[2]**Hanuman** Hindu monkey god

Kite

by Melvin Burgess

Melvin Burgess has written many novels for young people, including
Junk, about teenage addiction, which won the Carnegie Medal.

In *Kite*, Taylor Mase's father is a gamekeeper, employed by the
domineering and cruel Reg Harris. The Mase family is dependent on
Harris not only for their livelihood but also for their home. Teddy
Harris, Harris's elderly uncle, values nature and strongly opposes his
nephew's attitude to wildlife.

Gordon's Tower was a round stone tower built a hundred and fifty
years before. It was a folly; it had no use but to be looked at, and
looked from. If you climbed to the battlements on the top you
could see for miles. The top of the hill where the tower stood was
overgrown with brambles and young trees, but on the lower slopes
were full-grown trees. At the bottom of the hill, in a dip in the land,
where the crowns of the trees were hidden by the surrounding
wood, the kites had built their nest in a slender oak tree. It was the
first nest built by kites outside Wales in over thirty years.

Taylor stood at the foot of the slender grey trunk of the oak
tree. Even though he was alone he felt a thousand eyes on him:
Teddy Harris, Alan, and every bird in the woods. Everything
alive seemed to be judging him.

He reached out and touched the rough bark.

He thought: this tree of all trees. This place, these hands,
this boy. Above him sat a kite on her eggs. Even in Wales there
were only a handful of such nests. The tree was holy.

But Taylor had not come in praise. He felt sick with greed.
He stared up into the trembling green canopy. He tried to soak
everything in; he wanted to remember this moment forever. But
he was too scared to concentrate. The eyes in his imagination
disturbed him.

For a happy moment he thought he was going to be too
scared to do it, but then he thought: they're going to kill it anyway.

He was only following orders. He put a foot on a cleft in the trunk and began to climb.

Harris had found Taylor the day before as he walked home from school. He clicked his tongue sympathetically and put an arm on his shoulder as they walked together down the road. 'Pesky relatives!' he said. 'Pesky, pesky relatives! My uncle, your dad. What a pair!' He grinned and chuckled without amusement. 'Tell you what, though . . . ' He bent close to Taylor's ear. 'I'm a little worried about your father,' he said in a little worried voice. 'I'd hate my uncle to catch him at that nest, you see. He'd be out of a job; you'd be out of a house. It is my house you live in, after all.' He smiled and winked. 'It's better for you to do it, you know. Of course your dad just has your best interests at heart. But I'm thinking of him. Here . . . ' He took a ten shilling note out of his pocket and stuffed it into Taylor's pocket. 'Tomorrow afternoon. It's all clear. Uncle's away, your dad'll be busy. I'll see to that.'

Harris stood up, winked, and was gone before Taylor could say a word.

Taylor knew it wasn't right and he hated Harris all the more for going behind his father's back. But he knew at once he was going to do it anyway. And not for Harris, either, although the ten bob[1] would be handy. Taylor was doing it for himself.

He wanted those eggs more than anything else on earth.

As he climbed, Taylor had a picture in his mind of the moment his head would appear over the edge of the nest. It would be a page from a history book. There would be the great bird, caught in her secrecy. He would stare at her and she would start out of her nest like an eagle.

But of course the kite heard him coming. She left the nest almost as soon as he lifted off the ground and all he caught was the briefest of glimpses. A great shape disappeared silently

---

[1] **ten bob**   archaic slang meaning ten shillings

behind the branches as he turned his head. Taylor groaned in disappointment. He had missed the only chance he would ever have to see a live kite close up.

He continued to climb, pushing through the bright green leaves, until at last his dusty, pale face came up by the edge of the untidy nest and he looked down into the holy place.

Three perfect eggs.

Taylor reached out a trembling hand to touch them. They were still warm with the heat of the kite's body. They were alive.

For a few seconds, Taylor soaked up the warmth and stared at the beautiful chocolatey markings on the shell. Then he picked the eggs up and carefully nestled them one after the other inside his shirt. He put them next to his skin as if he had come to care for them. Then he began to climb down.

Taylor had stolen up to the nest, but now that it was done he wanted only to get away. He scrambled down the tree and jumped the last few feet to the ground. On his belly the eggs cracked together. He closed his hands loosely about them and began to run, jumping over logs and fallen branches, skidding on the damp forest earth. He tore along in a rage. His hands cradled the eggs, but every time he heard them crack together he was glad. He hated them for what they had made him do, and for what he was going to do. He'd have thrown them away if he could, but he loved them too, and he was unable to stop himself going through with it.

As the boy fled, half a mile above him the kite flew west. The last kites were shy creatures who had learned to abandon their nests at the slightest danger. She flew fast and never looked back. She would never return.

Taylor's mother was hanging out the washing when he came in, and he was able to sneak past the bright white sheets without her seeing him. Upstairs he half drew the curtains, not that anyone could see in, before he carefully took the treasure from inside his shirt. He hardly dared look. He had treated them like rubbish, running like that!

Somehow, two of the eggs had survived his run without any harm at all. The third had two craters in it, one above and one to the side. Taylor traced the marks with his finger. There was no wet. Although the shell had been crushed the membrane beneath it was unbroken.

Taylor tossed the egg to one side in disgust. It was useless for his purposes; he was only interested in the shell.

Inside each egg lay a little hidden figure. Though they were as still as stones, the eggs were alive. Three tiny hearts beat as fast as a trembling leaf. With a stethoscope,[2] Taylor would have been able to hear them. If he had held them to a bright light he would have been able to see their shadows, and watch how they trembled and moved inside. They had been brought to life by the warmth of their mother's breast, and kept alive this last half hour by the warmth of Taylor's skin. Now, as they lay on the quilt, a little breeze stirred the curtains and blew across them. The eggs began to cool. As they cooled, they began to die.

Taylor was so anxious it felt as if someone was mincing his insides. He walked round the room bent double, holding his stomach. Then he knelt by the bed in front of the eggs and screwed his fists into his eyes. He stayed there for nearly a minute trying to think of nothing, but he kept seeing Teddy Harris standing in the woods, with his passionate, damp eyes.

Rarest bird in the country! Only twenty-four left. One bad winter, a couple of egg collectors . . . and they could be gone forever.

Taylor was nearly weeping. But then he hissed to himself, 'Don't be so *stupid*. Someone would have killed the kite and taken the eggs anyway. I've probably done it a favour. It'll fly away back to Wales now.'

He went to the cupboard to get out his equipment.

---

[2]**stethescope**   an instrument used for listening to the heart beat or breathing

Taylor laid his tools on the table and turned on the little lamp. He took the first egg from the bed and sat down to work.

The main tool was a sharp hatpin his mum had given him. He carefully chipped away at the top of the egg with the pin until he had made a tiny hole. Then, ever so gently, he pushed the hatpin inside.

Unseen behind the hard shell, the tiny kite half opened its beak, but it lived in a liquid world and it died without a sound.

Frowning, Taylor began to stir the pin about. It was lumpy. That meant the chick had begun to grow. Blowing it would be difficult.

He spent nearly a minute stirring and poking with the pin, trying to break up the half-developed chick. He turned the egg upside down and made a tiny hole at the other end. In here, too, he stirred. Then he held the egg over a white china mug, pressed the hole to his lips, and began to blow.

A thick trickle of bloody muck oozed out of the hole. A vein appeared, doubled up. It got stuck. Taylor blew harder. The vein bulged, then ran out in a lump. He took another breath and blew again.

Something was stuck tight.

Taylor's hands had started shaking again. After all this, it had to work, it had to! He tried stirring with the pin again before another blow, but nothing moved. He tried again with the pin. He blew harder; a little more oozed out. He blew harder . . . harder . . .

Taylor was so tense that his fingers tightened without him noticing. The egg crumpled in his hand without warning and a mangled clot fell into the cracked white mug.

Taylor dropped the crushed shell and danced like an imp silently around the table. Why had he done this?

He was ruining everything!

There were two remaining eggs on the bed, their hearts still beating. One egg was ruined; that meant there was only one chance left to add a kite's egg to his collection.

But there was another way to go about it.

Taylor gave way to the thought that was at the back of his head the whole time. He could give the egg to Teddy Harris! He would hatch it, rear the chick, let it go. Then everything would be okay. There would be another kite in the world to make up for the one his dad had shot. Taylor would have saved the eggs, not ruined them. He wouldn't even have to say he had tried to blow them. It was the right thing to do!

He went to touch the eggs. They were noticeably cooler. He took the second egg over to the table and picked up the hatpin.

This time he spent several minutes stirring the insides about, trying to make it all as fine as he could. But although the pin was sharp at the end, the sides were round and couldn't really cut anything up. He tried stirring with the point at different depths into the egg and scraping gently against the inside of the shell. At last he was ready to blow it.

He lifted the egg to his lips and blew, and almost at once his fingers squeezed together and the egg burst in his hand.

Taylor held his head and wept.

After a long time he opened his eyes to stare at the two minced, half-grown forms lying in the bottom of the mug. They were far more than a network of veins; they had begun to form organs. He hadn't stood a chance.

He turned and glared at the third egg lying on the bed. He was furious with it. He rushed over and bounced violently on the bed so that the egg rolled right up close to his knees. What use was it to him if he couldn't blow it? He jumped again. The egg might as well be smashed! He bounced up and down, harder and harder. But at the last moment he caught himself. He felt the egg press lightly against his skin. If he lifted his hands, it would be all over.

Tenderly, Taylor lowered himself to one side of the egg and cupped it in his hand. The egg was quite cool now. It had been lying on the bed for half an hour. Taylor dearly wanted to destroy it, out of anger and also because it was proof of his crime. But in the end, he loved it too much.

Putting the egg in his pocket he ran downstairs and out into the garden. A chicken coop lay in the shade of a holly hedge. Inside was a broody hen sitting on her eggs, bringing on next year's layers. Taylor opened the coop. The hen looked nervously at him. She rolled her eyes and clucked softly, but she was far too much in love with her own eggs to move.

Gently, Taylor eased the kite egg underneath her. The hen shifted and then settled back down on it. As he closed the door she sank low onto the eggs and fluffed up her feathers.

Taylor stood outside and sniggered to himself. What a treat! He'd love to be around to see that hen if the egg hatched. A whacking great kite chick – like a sheep bringing up a wolf!

But of course the egg would never hatch. It was cracked, the chick inside was certainly dead. Even if it was alive the hen would tread on the damaged egg and break it. But he couldn't throw it away.

Taylor heard his mum in the kitchen. He ducked down behind the raspberry canes and sneaked away.

He ran out of the gate and down to the village to call for Alan. He wanted to tell him about his adventure, but he knew at once he never could. He'd just say he'd been looking for the kites but had seen nothing. No one must ever know what had happened that afternoon.

Further reading

If you enjoyed the text, you may like to read the whole book. You may also like Melvin Burgess' first book, *The Cry of the Wolf* (Puffin Books, 1995), a gripping tale about a threatened species. But beware, it does contain violent scenes. Burgess' other books for young people include *Bloodtide* (Simon Pulse, 2007), a story of revenge based on an Icelandic saga and set in London in 2200. It is the first in a series.

Please remember, collecting eggs from wild birds is illegal and the penalties are severe. Find out about protecting birds at http://www.rspb.org.uk.

The Play of Kes

by Barry Hines and Allan Stronach

The Play of Kes is adapted from the book *A Kestrel for a Knave* by Barry Hines (Penguin Classics, 2000) and focuses on the unhappy life of Billy Casper, who trains a kestrel from a chick. The bird becomes his world, providing an attachment that is denied to him at home.

Act 1, scene 6

Watching a kestrel.

Billy is sitting watching a kestrel that has been collecting food for its young, which are nesting in an old monastery wall. The farmer, who the field belongs to, appears.

FARMER   Hey. What are you doing?

BILLY   Nothing.

FARMER   Go on then. Don't you know this is private property?

BILLY   No! Can I get up to that kestrel's nest?

FARMER   What kestrel's nest?

BILLY   Up that wall.

FARMER   There's no nest up there, son.

BILLY   There is. I've just been watching it fly in.
Pause

FARMER   And what you going to do when you get up to it? Take all the eggs?

BILLY   There's no eggs in, they're young 'uns.

FARMER   Then there's nothing to get up for then is there?

BILLY   I just wanted to look, that's all.

FARMER   And you'd be looking from six feet under if you tried to climb up there.

BILLY   Can I just have a look from the bottom then? I've never found a hawk's nest before. That's where it is, look, in that hole in the side of that window.

FARMER   I know it is, it's nested here donkey's years now.

BILLY   Just think, and I never knew.

FARMER   There's not many that does.

BILLY   I've been watching them from across in the wood. You ought to have seen them. One of them was sat on that telegraph pole for ages. I was right underneath it, then I saw its mate. It came from miles away and started to hover just over there. Then it dived down behind that wall and came up with something in its claws. You ought to have seen it mister.

FARMER   I see it every day. It always sits on that pole.

BILLY   I wish I could see it every day. Has anybody ever been up that wall to look?

FARMER   Not that I know of. It's dangerous. They've been going to pull it down for ages.

BILLY   I bet I could get up.

FARMER   You're not going to have the chance though.

BILLY   If I lived here I'd get a young 'un and train it.

FARMER	Would you?
	Pause
BILLY	You can train them.
FARMER	And how would you go about it?
	Pause
BILLY	Do you know?
FARMER	No, and there's not many that does. That's why I won't let anyone near, because if they can't be kept properly it's criminal.
BILLY	Do you know anyone who's kept one?
FARMER	One or two. Not many.
BILLY	Where could you find out about them?
FARMER	Books I suppose. I should think there are books on falconry.[1]
BILLY	Think there'll be any in the library?
FARMER	Could be in the City library. They've books on everything there.
BILLY	I'm off now then. So long mister.
	Billy runs off
FARMER	Hey!
BILLY	What?
FARMER	Go through that gate. Not over the wall.

---

[1]**falconry**   a sport that involves trained birds of prey

Act 1, scene 12

The English lesson.

Mr Farthing is giving books out to the class when Billy walks in.

BILLY	I've just been to see Mr Gryce, sir.
MR FARTHING	Yes, I know. How many this time?
BILLY	Two.
MR FARTHING	Sting?
BILLY	Not bad.
MR FARTHING	Right, sit down then.

Pause as Billy sits down.

Right… We've been talking about fact and fiction. I want you to stand up and tell us something about yourself – a fact – that is really interesting.

BILLY	I don't know any sir.
MR FARTHING	I'm giving you two minutes to think of something lad, and if you haven't started then the whole class is coming back at four.

Immediate reaction from the class. Various pupils shout out.

HOLMES	Come on Billy.
CARTLEDGE	Or else you die!
BAKER	Say anything.
BARR	If I've to come back I'll kill him.
MR FARTHING	I'm waiting Casper.

Pause

BAKER	Tell him about your hawk Casper.

MR FARTHING	If anyone else calls out it will be the last call he'll make… What hawk Casper?… It is a stuffed one?
	The whole class laugh. Billy is upset and he wipes his eyes.
	What's so funny about that? Well Tibbut?
TIBBUT	He's got a hawk sir, it's a kestrel. He's mad about it. He never comes out with anybody else now, he just looks after this hawk. He's crackers with it.
BILLY	It's better than you any day Tibbut.
	Mr Farthing sits down. Pause.
MR FARTHING	Now then Billy, come on, tell me about this hawk… where did you get it from?
	Billy is looking down at his desk.
BILLY	Found it.
MR FARTHING	Where?
BILLY	In a wood.
MR FARTHING	What had happened to it; was it injured?
BILLY	It was a young one. It must have tumbled from a nest.
MR FARTHING	And where do you keep it?
BILLY	In our shed.
MR FARTHING	Isn't that cruel?
	Billy looks at him for the first time.
BILLY	I don't keep it in the shed all the time. I fly it every day.

MR FARTHING	And doesn't it fly away? I thought hawks were wild.
BILLY	'Course it doesn't fly away. I've trained it.
MR FARTHING	Was it difficult?
BILLY	'Course it was. You've to be right … right patient with them and take your time.
MR FARTHING	Come out here then and tell us all about it. *Billy goes out hesitatingly.* Right, how did you set about training it?
BILLY	I started training Kes when I'd had her about a fortnight. She was as fat as a pig though at first. You can't do much with them until you've got their weight down. Gradually you cut their food down, until you go in one time and they're keen. I could tell with Kes because she jumped straight on my glove as I held it towards her. So while she was feeding I got hold of her jesses.
MR FARTHING	Her what?
BILLY	Jesses. She wears them on her legs all the time so you can get hold of them as she sits on your glove.
MR FARTHING	And how do you spell that?
BILLY	J-E-S-S-E-S.
MR FARTHING	Right, tell us more.
BILLY	Then when she's on your glove you get the swivel – like a swivel on a dog lead – then you

thread your leash – that's a leather thong –
through your swivel, do you see?

MR FARTHING   Yes, I see. Carry on.

BILLY   So you wrap your leash round your fingers so
Kes is now fastened to your hand. When you've
reached this stage and she's feeding from your
hand regular and not bating too much…

MR FARTHING   Bating … what's that?

BILLY   Trying to fly off, in a panic like. So now you
can try feeding her outside and getting her
used to things.

Billy is now becoming more confident in telling his story.

But you start inside first, making her jump on
to your glove for the meat. Every time she
comes you give her a scrap of meat. A reward
like. Then when she'll come about a leash
length straight away you can try her outside,
off a fence or something. You put her down,
hold on to the end of the leash with your right
hand and hold your glove out for her to fly to.

*Billy is now doing the basic mime actions to accompany the
story.*

When she's done this a bit you attach a creance
instead of a leash – that's a long line, I used a
fishing line. Then you put the hawk down on
the fence post. Then you walk into the middle
of the field unwinding the creance and the
hawk's waiting for you to stop and hold your
glove up. It's so it can't fly away you see.

MR FARTHING   It sounds very skilful and complicated Billy.

BILLY   It doesn't sound half as bad as it is though. I've told you in a couple of minutes but it takes weeks to do all that. They're as stubborn as mules, hawks. Sometimes she'd be all right, then next time I'd go in the shed and she'd go mad, screaming and bating as though she'd never seen me before. You'd think that you'd learnt her something, you'd put her away feeling champion and then the next time you went you were back where you started.

MR FARTHING   You make it sound very exciting though.

BILLY   It is, but the most exciting thing was when I flew her free for the first time. You ought to have been there then. I was frightened to death.

Mr Farthing turns to the class.

David Bradley as Billy Casper from *Kes*.

MR FARTHING	Do you want to hear about it?
CLASS	Yes sir.
MR FARTHING	Carry on Casper.
BILLY	Well, I'd been flying her on the creance for about a week and she was coming to me anything up to thirty, forty yards. It says in the book that when it's coming this far, straight away, it's ready to fly loose. I daren't though. I kept saying to myself. I'll just use the creance today to make sure, then I'll fly it free tomorrow. I did this for about four days and I got right mad with myself. So on the last day I didn't feed her up, just to make sure that she'd be sharp set the next morning. I hardly went to sleep that night, I was thinking about it that much. When I got up next morning – it was a Saturday – I thought right, if she flies off, she flies off and it can't be helped. So I went down to the shed. She was dead keen as well, walking about on her shelf behind the bars and screaming out when she saw me coming. So I took her out on the field and tried her on the creance first time and she came like a rocket. So I thought right, this time. I unclipped the creance and let her hop on to the fence post. There was nothing stopping her now. She could have flown off and there was nothing I could have done about it. I was terrified. I thought, she's forced to go, she's forced to go. She'll just fly off and that will be it. But she didn't. She just sat there looking round while I backed off into the

field. I went right into the middle. Then I held my glove up and shouted her.

He is miming the action.

Come on Kes, come on then. Nothing happened at first. Then just as I was going to walk back to her. she came. Straight as a die, about a yard off the floor. She came twice as fast as when she had the creance on. She came like lightning, head dead still and her wings never made a sound. Then wham! Straight on to the glove, claws out grabbing for the meat. I was that pleased I didn't know what to do with myself, so I thought, just to prove it, I'll try her again, and she came the second time just as good. Well that was it. I'd trained her. I'd done it.

MR FARTHING   Right, that was very good. I enjoyed that and I'm sure the class did as well.

Splatter of applause from the class, Billy sits down.

Further reading

Barry Hines' novel *A Kestrel for a Knave* (Penguin Classics, 2000) is a compelling story and certainly worth reading. Try to watch the film *Kes* (MGM) as well if you can.

If you are interested in the study of birds and their relationships with humans, *King Solomon's Ring* (Routledge, 2002), by the Austrian zoologist Konrad Lorenz, is a fascinating read, and includes his work with jackdaws.

The Fish

by Elizabeth Bishop

The American poet Elizabeth Bishop conjures up a dramatic poem
in which an old battered fish is caught at the end of a line.

I caught a tremendous fish
and held him beside the boat
half out of water, with my hook
fast in a corner of his mouth.
He didn't fight.
He hadn't fought at all.
He hung a grunting weight,
battered and venerable
and homely. Here and there
his brown skin hung in strips
like ancient wallpaper,
and its pattern of darker brown
was like wallpaper:
shapes like full-blown roses
stained and lost through age.
He was speckled with barnacles,
fine rosettes of lime,
and infested
with tiny white sea-lice,
and underneath two or three
rags of green weed hung down.
While his gills were breathing in
the terrible oxygen
– the frightening gills,
fresh and crisp with blood,
that can cut so badly –
I thought of the coarse white flesh
packed in like feathers,
the big bones and the little bones,

the dramatic reds and blacks
of his shiny entrails,
and the pink swim-bladder
like a big peony.
I looked into his eyes
which were far larger than mine
but shallower, and yellowed,
the irises backed and packed
with tarnished tinfoil
seen through the lenses
of old scratched isinglass.
They shifted a little, but not
to return my stare.
– It was more like the tipping
of an object toward the light.
I admired his sullen face,
the mechanism of his jaw,
and then I saw
that from his lower lip
– if you could call it a lip –
grim, wet, and weaponlike,
hung five old pieces of fish-line,
or four and a wire leader
with the swivel still attached,
with all their five big hooks
grown firmly in his mouth.
A green line, frayed at the end
where he broke it, two heavier lines,
and a fine black thread
still crimped from the strain and snap
when it broke and he got away.
Like medals with their ribbons
frayed and wavering,
a five-haired beard of wisdom
trailing from his aching jaw.

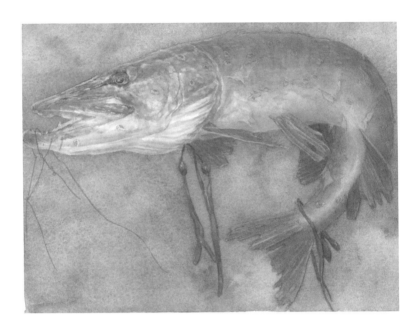

I stared and stared
and victory filled up
the little rented boat,
from the pool of bilge
where oil had spread a rainbow
around the rusted engine
to the bailer rusted orange,
the sun-cracked thwarts,
the oarlocks on their strings,
the gunnels – until everything
was rainbow, rainbow, rainbow!
And I let the fish go.

Further reading

The poet Marianne Moore has also written poems with vivid images of
fish and the sea. You might enjoy *The Fish* and *A Jelly-Fish*, published in
Complete Poems (Penguin Books, 1994).

Ted Hughes' famous poem *Pike*, published in *New Selected Poems
1957–1994* (Faber and Faber, 2001), may interest you as well.

The Sudden Knowledge of Moles
by Mike Gould

Mike Gould has written many books for young people, including educational books, short stories and drama. In this poem he describes an unexpected encounter with moles in a French garden.

Moles had been under my radar for years.
Since a half-forgotten camping trip
When our ground-sheet rumbled with
The dark uprooting of a pulsing ball,
They had waited, land-locked in black.
Now, in the garden at Saint Denis-le-Gast[1]
Their loamy hillocks erupt like pockets
Turned inside-out upon the green,
And the lawn's fair face is secretly corrupted.
Stubborn, lumpy clumps, the upward dumps
Defeat my fragile mower's sputter,
And later I'm seeing moles everywhere,
Or rather, *not* seeing them at all,
Save their abandoned traces – grassy tussocks,
In field and pastures beyond the railway's tracks.
I like that both of us were always there,
Paralleling worlds of air and dust.
While I traversed their earth's inverted crust,
They were workers in the weighty clay of time.
While I slept through the dawn's upheaval
They were nosing through their ocean's business,
Grubbing up and down their private furrows,
Perfect in their shadow prime.

---

[1]**Saint Denis-le-Gast**   pronounced *San Denee le Ga*

Further reading

Sylvia Plath's poem *Blue Moles*, which you can find in *Collected Poems* (Harper Perennial Modern Classics, 2008), has some startling images and looks at these underground creatures from a different point of view from *The Sudden Knowledge of Moles*. You could compare the two poems.

Wild Side of Town

by Chris Packham

Chris Packham is a naturalist and photographer as well as a television presenter. He has written several wildlife books, often related to urban wildlife. Here he studies the habitat needs of the Black Redstart.

Upright, twitchy and quick, in a dusty black coat with a brick red disappearing tail, the male Black Redstart is a great little bird and, above all, in Britain at least, it is the definitive urban species. That is to say, it doesn't occur in any of its primary natural habitats, only in our cities. A Robin-related member of the thrush family, it struggles here on the northern edge of its range, more at home on the warmer, insect-rich rocky slopes and screes of southern Europe. However, postwar bombsites, old docks and redundant industrial land physically replicate its favourite hunting and nesting habitats, and the extra few degrees of temperature above ambient in the city centres helps with winter warmth and keeps the insects moving longer and later. Despite our run of milder winters the population has fluctuated for the last 50 years and currently flits between 75 and 100 pairs, making it one of our rarest birds. Sadly, its future conservation will not be easy. Following the publication of the Urban White Paper in 2000, current development practices favour building on 'brownfield' rather than 'greenfield' sites, a policy that appears to suit most of our ideals of wildlife preservation. Also, because most of these areas which are Black Redstarts' breeding sites are deep within city centres, they represent priceless parcels of prime real estate. So is there a compromise that may give conservation a chance? Yes, and its champion is Mr D. Gedge.

Dusty Gedge is the lead for the Black Redstart Action Plan in London and regularly acts as a consultant to property developers, architects and various agencies. He is not one of the old school of old-fashioned single species conservationists; he has bigger ideas, namely 'Brownfield Biodiversity', and his ambitions are at the highest level, literally, because Dusty's solutions are on the roof.

On the continent, 'green' roofs are commonplace and often a statutory requirement in Germany and Switzerland. In Britain, as ever, we've been too conservative in our approach and aside from the top of the odd environmental centre you're unlikely ever to encounter the phenomenon. Some such roofs are genuinely green, planted with grass or sedum, but many are essentially roofs which have a substrate added to act as a sponge, typically crushed brick or concrete. Of course, such a surface is rapidly colonized by the local ruderals but often sparsely and thus the green roofs are in fact grey, which is not a handicap because this closely resembles the redstart's brownfield requirements. In Europe Black Redstarts or Biodiversity are not the most important issues for planners considering green roofs. Reduction of water run off as a flood protection is top of their agenda, followed by energy conservation through improved insulation and access to visual green space, which sees many green-roofed hospitals given the go-ahead. Dusty's drive to encourage the practice here has been frustrating as investors, planners and buyers are simply unfamiliar with the continental habit, and even other conservationists have displayed scant regard for his ideas. However, his perseverance is paying off; several schemes in the capital will be putting grass on a different level. Deptford Creek and Gargoyle Wharf will see extensive green roofs for the benefit of redstarts and others and, in fact, green roofing systems are going up where these illusive birds have never stretched their wings. Dusty is also dealing with architects and developers at an earlier stage as they become keen to incorporate eco-friendly features into the initial design and thus save costly changes at a later stage, for planners too are now recognizing brownfield ecological issues. English Nature is currently investigating the role green roofs can play in urban biodiversity and as more go up, and familiarity spreads, Dusty hopes that the roofs will cover urban Britain in time to secure a safe future for his favourite Black Redstarts.

Further reading

In *Wild Side of Town* (New Holland Publishers Ltd, 2007) you can find numerous accounts of city wildlife. It is divided into different chapters covering insects, species such as rats that adapt well to different conditions, birds, plant life, reptiles, amphibians and fish. It also looks at projects that support wildlife in the city and has a useful field guide at the end.

You could also visit http://www.bbc.co.uk/springwatch, the website of the BBC *Springwatch* series, where you can find out what an extraordinary range of wildlife lives in an ordinary urban garden.

Snakes

by A. K. Ramanujan

A. K. Ramanujan was an Indian author and scholar who studied several languages and published several volumes of poetry. In this verse he describes a child's response when the snakeman arrives with his ritual cobras.

A basketful of ritual cobras
comes into the tame little house,
their brown-wheat glisten ringed with ripples.
They lick the room with their bodies, curves
uncurling, writing a sibilant alphabet of panic
on my floor. Mother gives them milk
in saucers. She watches them suck
and bare the black-line design
etched on the brass of the saucer.
The snakeman wreathes their writhing
round his neck
for father's smiling
money. But I scream.

Further reading

You may like to read the whole of this poem, and also other poems by A. K. Ramanujan, such as *Self-Portrait*, an exploration of identity. You can find it in his *Collected Poems* (OUP India, 1997).

Another poem that depicts the snake, and the complex human responses it evokes, is *Snake* by D. H. Lawrence. This skilfully written, compelling poem can be found in *Complete Poems* by D. H. Lawrence (Penguin Books Ltd, 1994).

Life in Cold Blood

by David Attenborough

David Attenborough is a household name, largely because of his high-quality natural history programmes, which he has made for over fifty years. He was knighted in 1985. *Life in Cold Blood* was originally produced as a television series in 2008 and looks at the characteristics and evolution of reptiles and amphibians.

Efficient movers though they can be, pythons and boas seldom pursue their prey. They prefer to wait for it to come to them. They lie motionless, camouflaged by their skin patterns to the point of invisibility.

The absence of eyelids gives all snakes a fixed stare that suggests they are examining their surroundings with great intensity. In fact, boas and pythons are very short-sighted and can only focus their eyes to a limited extent. Nor have they the muscles needed to move their eyes in their sockets. So their eyesight is poor.

Their hearing is even worse. Their ancestors lost their eardrums and the passages that led to them when they took to burrowing, so now the snakes can only detect sound by resting their lower jaw on the ground. Vibrations in the earth such as those caused by a footstep can then be picked up and transmitted along the quadrate, the bone that connects the back of the jaw to the main part of the skull, and thus reach the capsule that contains the inner ear.

But though their hearing and sight are seriously limited, many pythons and boas have additional ways of detecting what goes on around them. The scales on the upper and lower margins of the mouth carry a line of small pits. These are sensitive to heat and can detect changes of a tiny fraction of a single degree Centigrade. That sense must undoubtedly have been of great value to the early snakes in their pursuit of the recently evolved small warm-blooded mammals.

And the snakes have one further sense organ – the tongue. Snake tongues are long and forked and similar in form to the tongues of monitor lizards – another reason for supposing that ancient monitors were the group that gave rise to the ancestral burrowing snakes in the first place. Snakes, unlike lizards, do not need to open their mouths to flick the tongue out, for they have developed a small notch at the front of the upper jaw through which the tongue can slide. And as they withdraw it, so it carries back molecules of scent floating in the air which are then wiped on to twin sense organs in the roof of their mouths, one for each prong of the fork.

Thus a python, lying motionless except for the silent flick of its tongue, may not be able to see an approaching rat with any clarity, but it can sense its presence from a considerable distance by its smell and accurately track its movements from the heat of its body even in the dark – which is usually the time when such small creatures are out foraging. The rat, however, has no way of detecting the snake. There is no movement, no conspicuous silhouette to warn it. And then suddenly the snake strikes. Its head shoots forward, jaws agape. The curved backward-pointing teeth fasten on the rat. Immediately, the python throws one or more coils around the rat and begins to squeeze its victim. As the muscles of its body contract, the coils tighten. Within seconds, the rat can no longer breathe. The pressure upon it increases so greatly that the heart itself is stopped. Few if any of the rat's bones have been broken. There was no need. It died within seconds of being seized.

Now the process of feeding begins. The parts of the snake's skull that lie below the stout, rigid box containing the brain are not fused together but only loosely connected by ligaments. Thus, the lower jaw, which in mammals is a single rigid bone, in a snake consists of two linked parts that can be moved independently. As the snake gapes to engulf its prey, the connecting ligaments stretch and the two halves of the lower jaw separate. There are other flexibilities too. Half way along the lower jawbone, on each side, there are joints. These allow the side of the

jaw to bow outwards. At the back, the quadrate bones that con-
nect the lower jaw to the upper part of the skull also hinge outwards. As a result of all these movements, the snake's gape can
become considerably wider than its head.

By moving each half of its lower jaw independently and
alternately the snake slowly changes its grip until its victim is
held head-first in its mouth. Prey, whatever it is, is usually
engulfed head-first so that it will be moved along and not
against the lie of fur, scales or feathers. Having placed the rat
correctly, the snake slowly works it down into its throat.
Swallowing may take several minutes and with its mouth
stuffed so full, the snake might have difficulty in breathing
during this long time. However it is able to force the upper end
of its windpipe out beside the body of its prey so that it can take
air into its lungs as it feeds.

The body of its victim travels slowly down the snake's
throat. The snake's lack of front limbs and pectoral girdle
means that there is no bony ring encircling its shoulders
through which its prey has to pass. The skin of the snake's
tubular body is also elastic and stretches as the muscles of its

body wall steadily force the meal down towards the stomach where the process of digestion will at last begin. If the meal has been a big one, this may take some time. If its victim had spines or even horns then sudden movement could cause a puncture of the snake's body wall. So the snake will now do its best to keep out of harm's way and avoid too much activity.

Particularly large meals stimulate changes in the snake's internal organs that are necessary to deal with the task of digestion and storage. Its heart swells by 40%. Within two days, its liver has doubled in size. Absorbing the whole meal may take a week or more. When at last the task is completed, the snake's bodily systems shut down once again, leaving only the equivalent of a pilot light activated.

The paradise flying snake (*Chrysopelea paradisi*) of the south-east Asian rainforest has become specialised in a way that parallels the gliding technique of the little flying lizard (*Draco volans*) that lives in the same trees. It is a small and particularly beautiful species with elegant golden lines across its head and yellow and scarlet scales scattered among the black ones that cover most of its body. It can climb vertical trunks at great speed using the standard serpentine method of pressing the curves of its body against roughnesses of the bark.

But it has an even more remarkable way of travelling. Having reached the canopy it may decide that it would prefer to look for prey in another tree. It races along a branch and, on reaching the end, launches itself into the air. As it leaves the branch, it maintains its lateral coils so that its silhouette approximates more to a squarish rectangle than a straight line. At the same time it flattens its body and pulls the margin of its flanks downwards so that its underside becomes visibly concave and traps the air beneath it, so slowing its fall. It is able to do this particularly effectively because the transverse scales on its underside have hinges at each end that allow their tips to be bent downwards. In this way a flying snake can easily travel twenty or thirty yards horizontally from its take-off point.

It can even, to some degree, steer while in the air. We once, in Borneo, started to film such a flight. Having captured a flying snake in the forest we took it to a water-tower in the forestry station where we were staying. In order to be sure of its exact take-off point and thus focus our cameras on it from the very beginning of its glide, we tied one end of a short plastic hose to the tower rail and then inserted the snake head-first into the other. *Chrysopelea* came out of the far end, executed a splendid glide and landed on the lawn of neatly clipped grass that surrounded the tower. We immediately caught it and took it back up the tower so that we might get a second rather tighter close-up shot. The same thing happened. So we took it back for a third flight. But this time – seemingly – it had assessed the situation. Somewhat to one side at the edge of the lawn there was a large and dense clump of very tall bamboo. As the little snake glided down for the third time it suddenly executed a neat left-hand turn and landed deep in the bamboo clump where we could not possibly re-catch it.

Further reading

Find out about the amphibian crisis that David Attenborough is concerned with. For example, go to http://www.bbc.co.uk/sn/tvradio/ and search for '"Last wave" for wild golden frog'.

David Attenborough is also concerned about the effects of global warming. In 2006 he presented two programmes as part of the BBC's *Climate Chaos* series: *Are We Changing Planet Earth?* and *Can We Save Planet Earth?* To find out more go to http://www.wikipedia.org and search for the first of these programmes.

Activities

My Life with the Chimpanzees

Before you read

1 Chimpanzees are closely related to humans. In what ways are they like us? Discuss your ideas with a partner, thinking about their expressions and behaviour, and draw up a list of similarities.

What's it about?

2 What words or phrases on page 2 tell you that:
   - there was excitement in the camp
   - the writer had been waiting a long time for the event
   - she was also anxious?

3 Draw a spider diagram with 'Chimps' in the middle. Add all the adult chimps mentioned in the text and write a few words to describe each one. Discuss with a partner how the chimps have different personalities.

4 Find the paragraph in the text that describes the chimps in the rain. In what ways is their behaviour like that of human beings? Are any of the similarities the same as the ones you listed in Activity 1?

Thinking about the text

5 Re-read the paragraphs in which the author has a terrifying encounter with the chimps (pages 4–5). Find some phrases that help to build tension, for example 'a chimp hunched in front of me' (page 4). Also find some examples of how tension is built up using short abrupt sentences. Then rewrite the description, but this time David Greybeard should be present. How would the chimps behave? How might the events change and the tension gradually disappear?

6 Jane Goodall kept a diary of everything the chimps did. Write her journal for three days. Include:
   - notes on the working conditions and weather
   - important observations about the chimps
   - what new information was learned.

7 Some scientists work in a different way from Jane Goodall. They believe that they should have limited contact when studying wild animals, to prevent becoming personally involved. Carry out a group discussion on the advantages and disadvantages of both approaches.

Maninagar Days

Before you read

1   Have you had a memorable encounter with a wild creature? What happened? What feelings did it evoke? How did the creature respond to you? Discuss your memory with a partner.

What's it about?

Read the text and make notes in answer to questions 2 to 4.

2   What is the setting of the poem? How is it depicted in the first two verses? Find words that tell you:
    ● what the climate is like
    ● in what part of the world it could be.

3   In what way are the Rhesus monkeys like us?

4   How do the children relate to the monkeys in the first verse?

Thinking about the text

5   The children's relationship with the monkeys is more complex than described in the first verse. Imagine what it might be like if monkeys lived in your garden. Think about:
    ● the familiarity of the monkeys
    ● the monkeys' intelligence
    ● the monkeys' playfulness.

    Discuss your ideas with a partner, then make notes on your own.

6   The poet appeals to a variety of senses that help to create the atmosphere of the garden. All appeal to sight, but what other senses do the following appeal to?
    a   the canna flower beds / still moist from the mornings watering
    b   screaming through / the trees
    c   the soft grey body

    Use three images of your own to describe monkeys in a garden or in the wild. Each should appeal to more than one sense.

7   The Hanuman monkey has 'lanky dancer's steps'. Write your own descriptive story about a captive monkey. Imagine that you regularly see it somewhere, for example at the zoo. Focus on its behaviour and appearance, contrasting its circumstances with those in *Maninagar Days*. Use vivid images to conjure up a picture of it.

Kite

Before you read

1   Think back to a time when you were small and you did something that you knew was wrong. Did you worry that others might find out or disapprove? Did you do anything to make amends? What are your feelings now as you look back? Share your ideas in a small group.

What's it about?

Answer questions 2 to 4 by yourself. Find quotations in the text that support your ideas.

2   What do we discover about kites in the opening paragraph? Why does this information make Taylor's theft even more disturbing?

3   Re-read the paragraph beginning 'Harris had found Taylor' (page 10). What do you think Harris has told Taylor to do? What is Taylor's attitude to him?

4   Why does Taylor want the eggs? What does he originally intend to do with them?

Thinking about the text

5   The paragraphs that deal with blowing the eggs are especially disturbing. Compare these to the paragraph in which the inside of the egg is described, beginning 'Inside each egg lay a little hidden figure' (page 12). Why do you think the author has written two such different accounts of the eggs? Discuss your ideas with a partner.

6   a   Find the paragraph in which Taylor remembers Teddy Harris' words. In what sense does Taylor have conflicting feelings about stealing the eggs? Discuss your ideas with a partner.
    b   Now carry out a hot-seating exercise with your partner. One of you should adopt the role of Taylor, who is in the hot seat. The other should adopt the role of Teddy Harris, who questions Taylor about his actions.

7   a   Draw up a two-column table headed 'Taylor'. In column 1, make notes about Taylor's character (e.g. determined). In column 2, add quotes to support what you say (e.g. 'he knew at once he was going to do it anyway' (page 10)).
    b   Write a character study of Taylor. Include one or two quotations.

The Play of Kes

Before you read

1 Birds of prey hunt almost entirely while flying. What birds of prey do you know that are similar to kestrels? What features do you think they have that make them efficient hunters? Note some examples.

What's it about?

Answer questions 2 to 4 by yourself. Find quotations in the text that support your ideas.

2 Where does Billy first see a kestrel? What do you learn about its habitat?

3 Billy uses several technical words in the text. Write them down and make notes on their meanings.

4 Read Mr Farthing's comment and the stage directions at the end of Scene 12. How does Mr Farthing respond to Billy's talk? How does the class respond? What do you think the class think of Billy? Why?

Thinking about the text

5 How does the farmer treat Billy when he first sees him and why? How do you think he regards Billy by the end of the scene? Write a paragraph to explain.

6 a Draw up two lists of adjectives and phrases. In the first, describe Billy before his performance. In the second, describe his actual performance.
   b Discuss your lists in a small group and together decide what effect the kestrel has had on Billy's desire to learn. In what way do you think the school has failed to do this for Billy?
   c Use the work you have done to write an essay called 'How to encourage true learning'.

7 Do some research and write an informative text. Choose
   ● an advice leaflet entitled 'What to do if you find a baby bird' *or*
   ● an account of falconry and different attitudes to it.

The Fish

Before you read

1   Work with a partner. One of you should describe a fish swimming and how we can see that water is its element (e.g. The silver body of the fish slithered through . . .). The other should describe a fish caught at the end of a line (e.g. Hooked and dangling in mid-air the fish seemed like . . .). Discuss the contrast.

What's it about?

Read the text and answer questions 2 to 4 by yourself. Then discuss your ideas with a partner.

2   What is the narrator's response to the fish when it is first caught? How does the fish react?

3   The narrator says 'I admired his sullen face' (page 27). What does 'sullen' mean? Why do you think the narrator admired the fish? With your partner, work out the meaning of these words from the context:

isinglass   thwarts   oarlocks   gunnel

Check your answers in a dictionary.

4   What happens in the last line of the poem and what do you think has led to this decision? Find evidence in the text that suggests this was always a possibility.

Thinking about the text

5   The fish is depicted in great detail. Search through the poem for all its features, making brief notes (e.g. brown skin hung in strips (page 26)).

6   The poet uses several similes to describe the fish, such as 'the coarse white flesh / packed in like feathers'.
   a   Find a simile that suggests the fish is like an old soldier who has won many battles.
   b   Find the other similes in the poem that conjure up the extreme age of the fish.

7   Write a poem or descriptive paragraph from the fish's point of view, remembering that he is experienced and wise. How does he feel? What does he see? What does he think the outcome will be for him? How does he react at the end? Use similes to create your images.

The Sudden Knowledge of Moles

Before you read

1   For us, unlike moles, being underground is not our natural element. But for some people caving or potholing is a great pleasure. What do you think the attractions are? Discuss your ideas with a partner.

What's it about?

Read the text and answer questions 2 to 4 by yourself. Write down your answers.

2   Why does the narrator of the poem say 'Moles had been under my radar for years'? What previous experience of moles has he had?

3   How do we know from the poem that moles are very difficult to see? Does the narrator see the moles or not?

4   Why do gardeners often dislike moles? Do you think the narrator feels the same way?

Thinking about the text

5   Write three paragraphs describing the setting of the poem, noting:
   - the type of countryside
   - notable features in the landscape
   - the effect the moles have.

6   The moles set the narrator thinking about the different ways humans and moles have evolved and their different habitats. Discuss with a partner what these lines mean:
   - Paralleling worlds of air and dust
   - While I traversed their earth's inverted crust
   - . . . workers in the weighty clay of time

7   **a**   In what way do all the words below suit the feelings expressed in the poem? Discuss your ideas with a partner.

   exasperation   affection   wonder   thoughtfulness

   Find lines in the poem to support your views.

   b   Write an essay entitled 'The feelings expressed in the poem'. Use the words from part a as a guide (e.g. you could write about the narrator's exasperation on finding the molehills) and use quotations to support what you say. You could begin like this: Different feelings are expressed in the poem *The Sudden Knowledge of Moles* . . .

Wild Side of Town

Before you read

1   In thirty seconds, think of as many birds as you can that inhabit our towns and cities and write them down. How many did you list?

What's it about?

Read the text and answer questions 2 to 4 by yourself.

2   Re-read the first two sentences. Why is the Black Redstart particularly interesting as an urban species?

3   Use context to work out the meaning of the following vocabulary.

habitat brownfield greenfield biodiversity sedum ruderals

Check your answers in a dictionary.

4   What does the author feel about British attitudes to conservation? Why does this make life for the Black Redstart more difficult?

Thinking about the text

5   **a**   Create a fact file for the Black Redstart. Write information under the following headings: Habitat, Food, Family, Rarity.
b   Draw an accurate picture of the Black Redstart, labelling its characteristics.

6   The author tells us that Dusty Gedge has big ideas. What are they? What does he mean by 'Brownfield Biodiversity' (page 32) and why is it important to the Black Redstart? Write an essay entitled 'A Solution for the Black Redstart'. Cover
   ●   the practice of green roofing in Europe and why it occurs
   ●   the problems encountered by Dusty Gedge
   ●   how these problems are being overcome by schemes in London
   ●   how these schemes would help the Black Redstart.

7   Does your school have a green roof in any part of the building?
   a   If so, write an account describing its function and how successful it is.
   b   If not, work in a small group and note down:
      ●   a suitable place for a green roof
      ●   how it might help wildlife and affect energy conservation
      ●   how much it might cost to develop one.

   Allot roles in the group and write up your findings.

Snakes

Before you read

1   Most snakes are not poisonous and those that are will only attack if provoked. They are timid creatures, frightened of humans, but we often have a fear of them. Think of several reasons why this is and discuss your ideas in a group.

What's it about?

2   Write a short paragraph about how the child feels about the snakes.

3   What do you think 'ritual cobras' are? Who do you think the snakeman is? Why might the snakes have been brought to the house? Write a paragraph offering several possibilities for the snakes' presence.

Thinking about the text

4   Study the way the poem is arranged. Some sentences stop in the middle of a line, creating a pause (caesura). Others carry on into the next line (run-on lines). The final line is short and stops abruptly. Read the poem aloud, noting these long and short lines. Then discuss in a group how these techniques help to suggest the movement of the snakes, or the child's fear.

5   **a**   What do you think the line 'writing a sibilant alphabet of panic' refers to? Think about the shapes made by the snakes and the child's attitude to the event. Then find two other vivid images of the snakes in the poem.
    b   Now create an image of your own for each of the following:
    ● a python
    ● a rattle snake.

6   In Pakistan and India *saperas* (snake charmers) appear to be able to control cobras through music. Use an encyclopaedia to check whether cobras are sensitive to music and carry out an investigation into cobras and *saperas*. Present a talk of your findings.

Life in Cold Blood

Before you read

1 David Attenborough points out in the Foreword to *Life in Cold Blood* that both reptiles and amphibians have evolved from some of the earliest species. What species do you think these are? Discuss your ideas in a small group and note some examples.

What's it about?

Read the text, then work with a partner to answer questions 2 to 4.

2 Snakes have poor eyesight and hearing. What various sense organs are especially acute and what do they use them for?

3 Re-read paragraphs 7 to 10 (from 'Now the process' (page 36) to 'a pilot light activated' (page 38)). Sum up how a python swallows its prey and digests it. How long might digestion of large prey take?

4 On page 39 the paradise flying snake 'seemingly . . . assessed the situation'. What does Attenborough mean by this?

Thinking about the text

5 a Paragraph 6 (beginning 'Thus a python' (page 36)) vividly describes a python catching its prey. The phrase 'motionless except for the silent flick of its tongue' adds drama to the first line. Find an example of repetition that also adds drama.

   b Now write a poem using the information in paragraph 6. Focus on:

   ● the python's agility
   ● the speed of the kill
   ● how disturbing it is from a human point of view.

6 How does the paradise flying snake 'fly' from tree to tree? Does it leap, glide or fly? Choose the correct method:

   ● It adopts a squarish shape and pulls its flanks downwards; the snake gains height and leaps from tree to tree.
   ● It adopts a squarish shape, then flattens its body sideways; the underside becomes concave, trapping air, and the snake glides.
   ● It has large scales on hinges that extend as flaps and which act as wings, so that it can fly.

   Check your answer is correct. Then write a full statement summing up how the paradise flying snake climbs trees.

Compare and contrast

1 Both *My Life with the Chimpanzees* and *Maninagar Days* focus on humans living with chimps and monkeys. Compare and contrast the experiences described, thinking about:
- attitudes to the chimps or monkeys
- how the animals behave
- how the animals are depicted by the writers.

Write an account, drawing on the texts to support your comments.

2 The extracts from *Kite* and *The Play of Kes* depict main characters who seem to have very different attitudes to birds of prey.
  a Work with a partner to discuss the differences between Taylor's treatment of the kites and their eggs, and Billy's treatment of Kes. Do you think there are any similarities in the two boys' attitudes?
  b Imagine the two boys meet. Take a role each and act out the conversation between the two boys about birds of prey.

3 The child in *Snakes* has an overwhelming fear of the cobras in the poem. David Attenborough, on the other hand, conveys his fascination and respect for reptiles. Write three short commentaries on snakes from the point of view of:
- a zoo keeper
- a conservationist
- the snake's prey (such as a mouse).

4 In what way have the texts in this section:
- broadened your factual knowledge of different animals
- confirmed your sympathy for or made you more sympathetic to animals
- made you want to read on or find out more?

Discuss your responses in a small group, referring to specific examples.

5 Which text would you recommend to:
- a friend of the same age as you
- your mother or father
- your teacher.

Write three short paragraphs explaining why.

2 Water worlds

This section looks at the essential element of life, water, which covers over two-thirds of the planet and has also shaped much of our landscape. We are fascinated by it and drawn to it. This is the common factor in the texts in this section. They explore water's magnetism, whether it be the sea and its ebb and flow in the poem *The Fringe of the Sea* or the pleasure of swimming and the feel of water in *Waterlog*. As you read the texts, think about why it is that, although we are land creatures, with a healthy regard for water's dangers, we are nonetheless intrigued by it.

Activities

1 **a** Work in a small group. Imagine the sea in a storm.
   - One member of the group thinks of a relevant word to describe what they see, such as 'squall'.
   - The next person responds immediately, saying a word or phrase that they associate with the first, such as 'buffeted' or 'high winds'.
   - Carry on until all the members of the group have contributed. Jot down the words as you go.
   - Continue until the group has a dozen memorable words.

   b Use the words to help you write a communal poem. You should agree on the style, how many verses to write and who will write what. Then bring your ideas together and shape your poem. (For example, the whole poem could be a series of images of ships at sea in a storm.)

2 The tides are the surface of the ocean rising and falling according to the moon and the sun's forces. A spring tide is unusually high and happens when the Earth, moon and sun are in line. Write a ghost story called 'Spring Tide'.

3 **a** Work in a small group. Find a map of the ocean currents in your school library or on the Internet (when using the Internet, always do so with your teacher's guidance). You will see that there are several important ones that can flow for thousands of miles.

b Find out about The Gulf Stream or The North Atlantic Drift. For example, go to http://www.howstuffworks.com and carry out a search. Then, in your group, prepare a factual account about this current and its importance on the climate. Include diagrams and captions.

c Present your work to another group.

Gone to Sea

by Michael Morpurgo

Michael Morpurgo has written numerous books for children over many years and was the Children's Laureate from 2003 to 2005. In this short story, which appears in *The White Horse of Zennor and other stories* (Egmont Books, 2008), William is the youngest of four brothers, living on a farm on the Cornish coast.

William Tregerthen had the look of a child who carried all the pain of the world on his hunched shoulders. But he had not always been like this. He is remembered by his mother as the happy, chortling child of his infancy, content to bask in his mother's warmth and secure in the knowledge that the world was made just for him. But with the ability to walk came the slow understanding that he walked differently from others and that this was to set him apart from everyone he loved. He found he could not run with his brothers through the high hay fields, chasing after rabbits; that he could not clamber with them down the rocks to the sea but had to wait at the top of the cliffs and watch them hop-scotching over the boulders and leaping in and out of the rock pools below.

He was the youngest of four brothers born onto a farm that hung precariously along the rugged cliffs below the Eagle's Nest. The few small square fields that made up the farm were spread, like a green patchwork between the granite farmhouse and the grey-grim sea, merging into gorse and bracken as they neared the cliff top. For a whole child it was a paradise of adventure and mystery, for the land was riddled with deserted tin miners' cottages and empty, ivy-clad chapels that had once been filled with boisterous hymns and sonorous[1] prayer. There were deserted wheel houses that loomed out of the mist, and dark, dank caves that must surely have been used by wreckers and smugglers. Perhaps they still were.

---

[1]**sonorous**   echoing

But William was not a whole child; his left foot was turned inwards and twisted. He shuffled along behind his older brothers in a desperate attempt to stay with them and to be part of their world. His brothers were not hard-souled children, but were merely wrapped in their own fantasies. They were pirates and smugglers and revenue men, and the shadowing presence of William was beginning already to encroach on their freedom of movement. As he grew older he was left further and further behind and they began to ignore him, and then to treat him as if he were not there. Finally, when William was just about school age, they rejected him outright for the first time. 'Go home to Mother,' they said. 'She'll look after you.'

William did not cry, for by now it came as no shock to him. He had already been accustomed to the aside remarks, the accusing fingers in the village and the assiduously[2] averted eyes. Even his own father, with whom he had romped and gambolled as an infant, was becoming estranged and would leave him behind more and more when he went out on the farm. There were fewer rides on the tractor these days, fewer invitations to ride up in front of him on his great shining horse. William knew that he had become a nuisance. What he could not know was that an inevitable guilt had soured his father who found he could no longer even look on his son's stumbling gait without a shudder of shame. He was not a cruel man by nature, but he did not want to have to be reminded continually of his own inadequacy as a father and as a man.

Only his mother stood by him and William loved her for it. With her he could forget his hideous foot that would never straighten and that caused him to lurch whenever he moved. They talked of the countries over the sea's end, beyond where the sky fell like a curtain on the horizon. From her he learned about the wild birds and the flowers. Together they would lie hidden in the bracken watching the foxes at play and counting the seals as they bobbed up and down at sea. It was rare enough

---

[2]**assiduously** tirelessly

for his mother to leave her kitchen but whenever she could she would take William out through the fields and clamber up onto a granite rock that rose from the soil below like an iceberg. From here they could look up to Zennor Quoit[3] above them and across the fields towards the sea. Here she would tell him all the stories of Zennor. Sitting beside her, his knees drawn up under his chin, he would bury himself in the mysteries of this wild place. He heard of mermaids, of witches, of legends as old as the rock itself and just as enduring. The bond between mother and son grew strong during these years; she would be there by his side wherever he went. She became the sole prop of William's life, his last link with happiness; and for his mother her last little son kept her soul singing in the midst of an endless drudgery.

For William Tregerthen, school was a nightmare of misery. Within his first week he was dubbed 'Limping Billy'. His brothers, who might have afforded some protection, avoided him and left him to the mercy of the mob. William did not hate his tormentors any more than he hated wasps in September; he just wished they would go away. But they did not. 'Limping Billy' was a source of infinite amusement that few could resist. Even the children William felt might have been friends to him were seduced into collaboration. Whenever they were tired of football or of tag or skipping, there was always 'Limping Billy' sitting by himself on the playground wall under the fuchsia hedge. William would see them coming and screw up his courage, turning on his thin smile of resignation that he hoped might soften their hearts. He continued to smile through the taunting and the teasing, through the limping competitions that they forced him to judge. He would nod appreciatively at their attempts to mimic the Hunchback of Notre Dame[4] and

---

[3]**Zennor Quoit**   ancient burial tomb (Quoit) of the village of Zennor in
   Penwith, Cornwall
[4]**Hunchback of Notre Dame**   the main character in a novel of the same
   name by French novelist Victor Hugo, published 1831

conceal his dread and his humiliation when they invited him to do better. He trained himself to laugh with them back at himself; it was his way of riding the punches.

His teachers were worse, cloaking their revulsion under a veneer of pity. To begin with they over-burdened him with a false sweetness and paid him far too much loving attention; and then because he found the words difficult to spell and his handwriting was uneven and awkward, they began to assume, as many do, that one unnatural limb somehow infects the whole and turns a cripple into an idiot. Very soon he was dismissed by his teachers as unreachable and ignored thereafter.

It did not help either that William was singularly unchild-like in his appearance. He had none of the cherubic innocence of a child; there was no charm about him, no redeeming feature. He was small for his age; but his face carried already the mark of years. His eyes were dark and deep-set, his features pinched and sallow. He walked with a stoop, dragging his foot behind him like a leaden weight. The world had taken him and shrivelled him up already. He looked permanently gaunt and hungry as he sat staring out of the classroom window at the heaving sea beyond the fields. A recluse was being born.

On his way back from school that last summer, William tried to avoid the road as much as possible. Meetings always became confrontations, and there was never anyone who wanted to walk home with him. He himself wanted less and less to be with people. Once into the fields and out of sight of the road he would break into a staggering, ugly run, swinging out his twisted foot, straining to throw it forward as far as it would go. He would time himself across the field that ran down from the road to the hay barn, and then throw himself at last face down and exhausted into the sweet warmth of new hay. He had done this for a few days already and, according to his counting, his time was improving with each run. But as he lay there now panting in the hay he heard someone clapping high up in the haystack behind him. He sat up quickly and looked around. It was a face he knew, as familiar to him as the rocks in the fields around the farm, an old face full of deeply etched crevasses and raised veins, unshaven and red with drink. Everyone around the village knew Sam, or 'Sam the Soak' as he was called, but no-one knew much about him. He lived alone in a cottage in the churchtown up behind the Tinners' Arms, cycling every day into St Ives where he kept a small fishing boat and a few lobster pots. He was a fair-weather fisherman, with a ramshackle boat that only went to sea when the weather was set fair. Whenever there were no fish or no lobsters to be found, or when the weather was blowing up, he would stay on shore and drink. It was rumoured there had been some great tragedy in his life before he came to live at Zennor, but he never spoke of it so no-one knew for certain.

'A fine run, Billy,' said Sam; his drooping eyes smiled gently. There was no sarcasm in his voice but rather a kind sincerity that William warmed to instantly.

'Better'n yesterday anyway,' William said.

'You should swim, dear lad,' Sam sat up and shook the hay out of his hair. He clambered down the haystack towards William, talking as he came. 'If I had a foot like that, dear lad, I'd swim. You'd be fine in the water, swim like the seals I shouldn't wonder.' He smiled awkwardly and ruffled William's hair. 'Got a lot to do. Hope you didn't mind my sleeping awhile in your hay. Your father makes good hay, I've always said that. Well, I can't stand here chatting with you, got a lot to do. And, by the by dear lad, I shouldn't like you to think that I was drunk.' He looked hard down at William and tweaked his ear. 'You're too young to know but there's worse things can happen to a man than a twisted foot, Billy, dear lad. I drink enough, but it's just enough and no more. Now you do as I say, go swimming. Once in the water you'll be the equal of anyone.'

'But I can't swim,' said William. 'My brothers can but I never learnt to. It's difficult for me to get down on the rocks.'

'Dear lad,' said Sam, brushing off his coat. 'If you can run with a foot like that, then you can most certainly swim. Mark my words, dear lad; I may look like an old soak – I know what they call me – but drink in moderation inspires great wisdom. Do as I say, get down to the sea and swim.'

William went down to the sea in secret that afternoon because he knew his mother would worry. Worse than that, she might try to stop him from going if she thought it was dangerous. She was busy in the kitchen so he said simply that he would make his own way across the fields to their rock and watch the kestrel they had seen the day before floating on the warm air high above the bracken. He had been to the seashore before of course, but always accompanied by his mother who had helped him down the cliff path to the beach below.

Swimming in the sea was forbidden. It was a family edict, and one observed by all the farming families around, whose respect and fear of the sea had been inculcated into them for generations. 'The sea is for fish,' his father had warned them often enough. 'Swim in the rock pools all you want, but don't go swimming in the sea.'

With his brothers and his father making hay in the high field by the chapel William knew there was little enough chance of his being discovered. He did indeed pause for a rest on the rock to look skywards for the kestrel, and this somehow eased his conscience. Certainly there was a great deal he had not told his mother, but he had never deliberately deceived her before this. He felt however such a strong compulsion to follow Sam's advice that he soon left the rock behind him and made for the cliff path. He was now further from home than he had ever been on his own before.

The cliff path was tortuous, difficult enough for anyone to negotiate with two good feet, but William managed well enough using a stick as a crutch to help him over the streams that tumbled down fern-green gorges to the sea below. At times he had to go down on all fours to be sure he would not slip. As he clambered up along the path to the first headland, he turned and looked back along the coast towards Zennor Head, breathing in the wind from the sea. A sudden wild feeling of exuberance and elation came over him so that he felt somehow liberated and at one with the world. He cupped his hands to his mouth and shouted to a tanker that was cruising motionless far out to sea:

'I'm Limping Billy Tregerthen,' he bellowed, 'and I'm going to swim. I'm going to swim in the sea. I can see you but you can't see me. Look out fish, here I come. Look out seals, here I come. I'm Limping Billy Tregerthen and I'm going to swim.'

So William came at last to Trevail Cliffs where the rocks step out into the sea but even at low tide never so far as to join the island. The island where the seals come lies some way off the shore, a black bastion against the sea, warning it that it must not come any further. Cormorants and shags[5] perched on the island like sinister sentries and below them William saw the seals basking in the sun on the rocks. The path down to the beach was treacherous and William knew it. For the first time he had to manage on his own, so he sat down and bumped his way down the track to the beach.

He went first to the place his brothers had learnt to swim, a great green bowl of sea water left behind in the rocks by the tide. As he clambered laboriously over the limpet-covered rocks towards the pool, he remembered how he had sat alone high on the cliff top above and watched his brothers and his father diving and splashing in the pool below, and how his heart had filled with envy and longing. 'You sit there, with your Mother,' his father had said. 'It's too dangerous for you out there on those rocks. Too dangerous.'

'And here I am,' said William aloud as he stepped gingerly forward onto the next rock, reaching for a hand-hold to support himself. 'Here I am, leaping from rock to rock like a goat. If only they could see me now.'

He hauled himself up over the last lip of rock and there at last was the pool down below him, with the sea lapping in gently at one end. Here for the first time William began to be

---

[5]**cormorants and shags** large diving seabirds

frightened. Until this moment he had not fully understood the step he was about to take. It was as if he had woken suddenly from a dream: the meeting with Sam in the hay barn, his triumphant walk along the cliff path, and the long rock climb to the pool. But now as he looked around him he saw he was surrounded entirely by sea and stranded on the rocks a great distance out from the beach. He began to doubt if he could ever get back; and had it not been for the seal William would most certainly have turned and gone back home.

The seal surfaced silently into the pool from nowhere. William crouched down slowly so as not to alarm him and watched. He had never been this close to a seal. He had seen them often enough lying out on the rocks on the island like great grey cucumbers and had spotted their shining heads floating out at sea. But now he was so close he could see that the seal was looking directly at him out of sad, soulful eyes. He had never noticed before that seals had whiskers. William watched for a while and then spoke. It seemed rude not to.

'You're in my pool,' he said. 'I don't mind really, though I was going to have a swim. Tell you the truth, I was having second thoughts anyway, about the swimming I mean. It's all right for you, you're born to it. I mean you don't find getting around on land that easy, do you? Well nor do I. And that's why Sam told me to go and learn to swim, said I'd swim like a seal one day. But I'm a bit frightened, see. I don't know if I can, not with my foot.'

The seal had vanished as he was speaking, so William lowered himself carefully step by step down towards the edge of the pool. The water was clear to the bottom, but there was no sign of the seal. William found it reassuring to be able to see the bottom, a great slab of rock that fell away towards the opening to the sea. He could see now why his brothers had come here to learn, for one end of the pool was shallow enough to paddle whilst the other was so deep that the bottom was scarcely visible.

William undressed quickly and stepped into the pool, feeling for the rocks below with his toes. He drew back at the first touch because the water stung him with cold, but soon he had both feet in the water. He looked down to be sure of his footing, watching his feet move forward slowly out into the pool until he was waist-high. The cold had taken the breath from his body and he was tempted to turn around at once and get out. But he steeled himself, raised his hands above his head, sucked in his breath and inched his way forward. His feet seemed suddenly strange to him, apart from him almost and he wriggled his toes to be sure that they were still attached to him. It was then that he noticed that they had changed. They had turned white, dead white; and as William gazed down he saw that his left foot was no longer twisted. For the first time in his life his feet stood parallel. He was about to bend down to try to touch his feet, for he knew his eyes must surely be deceiving him, when the seal reappeared only a few feet away in the middle of the pool. This time the seal gazed at him only for a few brief moments and then began a series of water acrobatics that soon had William laughing and clapping with joy. He would dive, roll and twist, disappear for a few seconds and then materialise somewhere else. He circled William, turning over on his back and rolling, powering his way to the end of the pool before flopping over on his front and aiming straight for William like a torpedo, just under the surface. It was a display of comic elegance, of easy power. But to William it was more than this, it became an invitation he found he could not refuse.

The seal had settled again in the centre of the pool, his great wide eyes beckoning. William never even waited for the water to stop churning but launched himself out into the water. He sank of course, but he had not expected not to. He kicked out with his legs and flailed his arms wildly in a supreme effort to regain the surface. He had sense enough to keep his mouth closed but his eyes were wide open and he saw through the green that the seal was swimming alongside him, close enough to touch. William knew that he was not drowning, that the seal would not let him drown; and with that confidence his arms and legs began to move more easily through the water. A few rhythmic strokes up towards the light and he found the air his lungs had been craving for. But the seal was nowhere to be seen. William struck out across to the rocks on the far side of the pool quite confident that the seal was still close by. Swimming came to William that day as it does to a dog. He found in that one afternoon the confidence to master the water. The seal however never reappeared, but William swam on now by himself until the water chilled his bones, seeking everywhere for the seal and calling for him. He thought of venturing out into the open ocean but thought better of it when he saw the swell outside the pool. He vowed he would come again, every day, until he found his seal.

William lay on the rocks above the pool, his eyes closed against the glare of the evening sun off the water, his heart still beating fast from the exertion of his swim. He lay like this, turning from time to time until he was dry all over. Occasionally he would laugh out loud in joyous celebration of the first triumph of his life. Out on the seal island the cormorants and shags were startled and lifted off the rocks to make for the fishing grounds out to sea, and the colony of seals was gathering as it always did each evening.

As William made his way back along the cliff path and up across the fields towards home he could hear behind him the soft hooting sound of the seals as they welcomed each new arrival on the rocks. His foot was indeed still twisted, but he walked erect now, the stoop gone from his shoulders and there was a new lightness in his step.

He broke the news to his family at supper that evening, dropped it like a bomb and it had just the effect he had expected and hoped for. They stopped eating and there was a long heavy silence whilst they looked at each other in stunned amazement.

'What did you say, Billy?' said his father sternly, putting down his knife and fork.

'I've been swimming with a seal,' William said, 'and I learnt to swim just like Sam said. I climbed down to the rocks and I swam in the pool with the seal. I know we mustn't swim in the sea but the pool's all right isn't it?'

'By yourself, Billy?' said his mother, who had turned quite pale. 'You shouldn't have, you know, not by yourself. I could have gone with you.'

'It was all right, Mother,' William smiled up at her. 'The seal looked after me. I couldn't have drowned, not with him there.'

Up to that point it had all been predictable, but then his brothers began to laugh, spluttering about what a good tale it was and how they had actually believed him for a moment; and when William insisted that he could swim now, and that the seal had helped him, his father lost his patience. 'It's bad enough your going off on your own without telling your mother, but then you come back with a fantastic story like that and expect me to believe it. I'm not stupid lad. I know you can't climb over those rocks with a foot like that; and as for swimming and seals, well it's a nice story, but a story's a story, so let's hear no more of it.'

'But he was only exaggerating, dear,' said William's mother. 'He didn't mean … '

'I know what he meant,' said his father. 'And it's your fault, like as not, telling him all these wild stories and putting strange ideas in his head.'

William looked at his mother in total disbelief, numbed by the realisation that she too doubted him. She smiled sympathetically at him and came over to stroke his head.

'He's just exaggerating a bit, aren't you Billy?' she said gently.

But William pulled away from her embrace, hurt by her lack of faith.

'I don't care if you don't believe me,' he said, his eyes filling with tears. 'I know what happened. I can swim I tell you, and one day I'll swim away from here and never come back. I hate you, I hate you all.'

His defiance was punished immediately. He was sent up to his room and as he passed his father's chair he was cuffed roundly on the ear for good measure. That evening, as he lay on his bed in his pyjamas listening to the remorseless ker-thump, ker-thump of the haybaler outside in the fields, William made up his mind to leave home.

His mother came up with some cocoa later on as she always did, but he pretended to be asleep, even when she leant over and kissed him gently on the forehead.

'Don't be unhappy, Billy,' she said. 'I believe you, I really do.'

He was tempted at that moment to wake and to call the whole plan off, but resentment was still burning too strongly inside him. When it mattered she had not believed him, and even now he knew she was merely trying to console him. There could be no going back. He lay still and tried to contain the tears inside his eyes.

Every afternoon after school that week William went back down to the beach to swim. One of his brothers must have said something for word had gone round at school that 'Limping Billy' claimed that he had been swimming with the seals. He endured the barbed ridicule more patiently than ever because he knew that it would soon be over and he would never again have to face their quips and jibes, their crooked smiles.

The sea was the haven he longed for each day. The family were far too busy making hay to notice where he was and he was never to speak of it again to any of them. To start with he kept to the green pool in the rocks. Every afternoon his seal would be there waiting for him and the lesson would begin. He learnt to roll in the water like a seal and to dive deep exploring the bottom of the pool for over a minute before surfacing for air. The seal teased him in the water, enticing him to chase, allowing William to come just so close before whisking away out of reach again. He learnt to lie on the water to rest as if he were on a bed, confident that his body would always float, that the water would always hold it up. Each day brought him new technique and new power in his legs and arms. Gradually the seal would let him come closer until one afternoon just before he left the pool William reached out slowly and stroked the seal on his side. It was gesture of love and thanks. The seal made no immediate attempt to move away but turned slowly in the water and let out a curious groan of acceptance before diving away out of the pool and into the open sea. As he watched him swim away, William was sure at last of his place in the world.

With the sea still calm next day William left the sanctuary of the pool and swam out into the swell of the ocean with the seal alongside him. There to welcome them as they neared the island were the bobbing heads of the entire seal colony. When they swam too fast for him it seemed the easiest, most natural thing in the world to throw his arms around the seal and hold on, riding him over the waves out towards the island. Once there he lay out on the rocks with them and was minutely inspected by each member of the colony. They came one by one and lay beside him, eyeing him wistfully before lumbering off to make room for the next. Each of them was different and he found he could tell at once the old from the young and the female from the male. Later, sitting cross-legged on the rocks and surrounded entirely by the inquisitive seals, William tasted raw fish for the first time, pulling away the flesh with his teeth as if he had been doing it all his life. He began to murmur seal noises in an attempt to thank them for their gift and they responded with great hoots of excitement and affection. By the time he was escorted back to the safety of the shore he could no longer doubt that he was one of them.

The notepad he left behind on his bed the next afternoon read simply: 'Gone to sea, where I belong.' His mother found it that evening when she came in from the fields at dusk. The Coastguard and the villagers were alerted and the search began. They searched the cliffs and the sea shore from Zennor Head to Wicca Pool[6] and beyond, but in vain. An air-sea rescue

---

[6]**Wicca Pool**   a cove off the Cornish Coast

helicopter flew low over the coast until the darkness drove it away. But the family returned to the search at first light and it was William's father who found the bundle of clothes hidden in the rocks below Trevail Cliffs. The pain was deep enough already, so he decided to tell no one of his discovery, but buried them himself in a corner of the cornfield below the chapel. He wept as he did so, as much out of remorse as for his son's lost life.

Some weeks later they held a memorial service in the church, attended by everyone in the village except Sam whom no one had seen since William's disappearance. The Parochial[7] Church Council was inspired to offer a space on the church wall for a memorial tablet for William, and they offered to finance it themselves. It should be left to the family they said, to word it as they wished.

Months later Sam was hauling in his nets off Wicca Pool. The fishing had been poor and he expected his nets to be empty once again. But as he began hauling it was clear he had struck it rich and his heart rose in anticipation of a full catch at last. It took all his strength to pull the net up through the water. His arms ached as he strained to find the reserves he would need to haul it in. He had stopped hauling for a moment to regain his strength, his feet braced on the deck against the pitch and toss of the boat, when he heard a voice behind him.

'Sam,' it said, quietly.

He turned instantly, a chill of fear creeping up his spine. It was William Tregerthen, his head and shoulders showing above the gunwale of the boat.

'Billy?' said Sam. 'Billy Tregerthen? Is it you, dear lad? Are you real, Billy? Is it really you?' William smiled at him to reassure him. 'I've not had a drink since the day you died,

---

[7]**Parochial**   belonging to a parish, local

Billy, honest I haven't. Told myself I never would again, not after what I did to you.' He screwed up his eyes. 'No,' he said, 'I must be dreaming. You're dead and drowned. I know you are.'

'I'm not dead and I'm not drowned, Sam,' William said. 'I'm living with the seals where I belong. You were right, Sam, right all along. I can swim like a seal, and I live like a seal. You can't limp in the water, Sam.'

'Are you really alive, dear lad?' said Sam. 'After all this time? You weren't drowned like they said?'

'I'm alive, Sam, and I want you to let your nets down,' William said. 'There's one of my seals caught up in it and there's no fish there I promise you. Let them down, Sam, please, before you hurt him.'

Sam let the nets go gently hand over hand until the weight was gone.

'Thank you Sam,' said William. 'You're a kind man, the only kind man I've ever known. Will you do something more for me?' Sam nodded, quite unable to speak any more. 'Will you tell my mother that I'm happy and well, that all her stories were true, and that she must never be sad. Tell her all is well with me. Promise?'

'Course,' Sam whispered. 'Course I will, dear lad.'

And then as suddenly as he had appeared, William was gone. Sam called out to him again and again. He wanted confirmation, he wanted to be sure his eyes had not been deceiving him. But the sea around him was empty and he never saw him again.

William's mother was feeding the hens as she did every morning after the men had left the house. She saw Sam coming down the lane towards the house and turned away. It would be more sympathy and she had had enough of that since William died. But Sam called after her and so she had to turn to face him. They spoke together for only a few minutes, Sam with his hands on her shoulders, before they parted leaving her alone again with her hens clucking impatiently around her feet. If

Sam had turned as he walked away he would have seen that she was smiling through her tears.

The inscription on the tablet in the church reads:

WILLIAM TREGERTHEN

AGED 10

Gone to sea, where he belongs

Further reading

If you enjoyed *Gone to Sea* you may also enjoy *Why the Whales Came* (Egmont Books Ltd, 2007), set on the Isles of Scilly in 1914. It was made into a film, *When the Whales Came*, in 1989.

Michael Morpurgo has written several other short stories. Read more from *The White Horse of Zennor* – such as the ghostly tale *The Giant's Necklace*.

Daughter of the Sea
by Berlie Doherty

Berlie Doherty's *Daughter of the Sea* (Andersen Press Ltd, 2008) tells the tale of a fisherman and his wife. When he rescues a creature from the sea and brings it home, he sets in motion a train of events that will profoundly change their lives. This is the preface to the book and the opening chapter.

Prologue

My tale is of the sea. It takes place in the far north, where ice has broken land into jagged rocks, and where black and fierce tides wash the shores. Hail is flung far on lashing winds, and winters are long and dark. Men haunt the sea, and the sea gives up to them a glittering harvest. And it is said that the people of the sea haunt the land.

My tale is of the daughter of the sea. The best way to hear the tale is to creep into the lee of the rocks when the herring boats have just landed. The gulls will be keening around you. The women hone knives on the stones, and their hands will be brown from the wind and the fish-gut slime. And as they work they talk to each other of things they've always known.

That's when the story's told.

Imagine a woman called Jannet, standing on the weed-wet stones. It would be dark, and the spray would be scraping her cheeks and the wind would be delving into her hair. She would be looking into the damsony dark and seeing nothing. And imagine her husband, Munroe Jaffray, crouching into his boat with the wild waves lumbering round him. And there's another to think of. Eilean o da Freya. Some say she's as weak in the head as a stunned herring. Others say she has the wisdom of the ancients. Jannet, Munroe, and Eilean. They're the ones who know for sure what happened on the night of the freak storm.

This is the tale.

The freak storm

Jannet Jaffray had gone to bed early and woke to the sound of a wild kind of wailing, far out and away at sea. It faded as soon as she sat up to listen. She lay in the dark hearing only the waves and the wind and this shadow of singing, and then put out her fingers to touch her husband's arm. With a missing beat of her heart she knew that Munroe was not home.

In minutes she was down on the shore, her lantern high in her hand, and counting the fishing boats that had been beached up by the rocks. As her lamp guttered its last breath she had counted all home except for Munroe's. The sea was high and frenzied; the moon was a ghost face, hiding and squinting as the dark clouds rolled.

'Munroe!' she called uselessly. It was on nights like this that the sea snatched lives instead of yielding fish. Jannet heard footsteps treading the sand behind her, and turned, smiling, expecting it to be her husband come home safe to her from another bay along the coast. There was no one to be seen. Yet again she heard footsteps in the shadows.

'Are you there?'

Nothing to hear or see.

'Then is it Eilean?'

She was answered this time by low singing coming like nightmares out of the huddled rocks. She knew the sound well.

'Go home, Eilean. There's nothing for you here.'

The woman Eilean edged closer, wet skirt clinging and hair like weeds round her face. She often came down to the shore on nights like this. It was as if she was searching for something that could never be found. She stood by Jannet's side, saying nothing.

Jannet stared out into the foam-flecked mouths of the waves. Munroe was somewhere there, and she would not be going home herself until she had sight of him.

Munroe was not far away. His boat was caught between the dangerous rocks of the skerries. He knew he must wait till first light before he tried to pull clear. The night had begun with a calm sea, though the fishing had not been good. He was about to turn for home when he heard a strange keening on the wind, swelling and fading like the flow of the tide itself. He saw that the water beneath his boat flashed with a streaming shoal of fishes. He followed it, leaning out in his eagerness to pull in the catch.

That was when the storm had risen out of nowhere. Waves snapped around him like a pack of wild dogs, forcing him into the ring of rocks for shelter. He was safe enough there. Just outside the skerries[1] there was a wild frenzy of waves that would toss his boat like a shell.

---

[1]**skerries**   a small rocky reef or island

Black clouds billowed across the sky. The moonlight came through in flashes, giving him brief glimpses of the waves and the tower of the spray. No going home yet, he thought.

It was in one of these moon flashes that he saw the child in the water.

It could be that Jannet fell asleep standing up that night, with the wind holding her round like ropes as she swayed. She didn't know that the tide had dropped or the sky was bleaching to grey until a voice called out to her: 'Jannet!' And there he was, riding high through the sea-mist – Munroe, her husband, safe home, and the small boat all in one piece. She ran knee-deep into the water to haul the boat in, and Munroe jumped out to help her.

'Why were you out there in this weather?' she shouted, the wind tearing her voice this way and that. He did not hear her, and she did not repeat it. He was safe, after all.

'I've a fine catch for you,' he told her. He reached back into his boat for the basket he used for carrying small fry. 'I'll show you when we get home.'

And just as they were leaving, with the boat pulled up on high sand and lashed to boulders, Eilean crept out of her shelter of rocks.

'I forgot about her!' Jannet whispered. 'She's been idling here all night waiting for your corpse to be washed ashore!'

The woman scuttled over to them, more like a crab than anything, the way she picked over the sharp pebbles. She tried to grab at the basket but Munroe swung it up away from her, holding it high above her head.

'Not for you!' he warned.

'What is it you have in there?' she asked him. 'A fish that sings. Is that it?'

Further reading

Berlie Doherty has a compelling, often lyrical style and if this appeals you may like to read the rest of the book. However, the author covers a wide range of issues. For example, you could find out how she addresses the issue of conservation by looking at *Tilly Mint and the Dodo* (Mammoth, 1996), which was written for younger readers and is concerned with species extinction. *Dear Nobody* (Puffin Books, 2001), for which she won the Carnegie Medal, is probably her best-known book; it deals with teenage pregnancy.

The Fringe of the Sea

by A. L. Hendriks

This poem by A. L. Hendriks explores another perspective of the sea and the profound effect it can have on the heart and mind of an islander.

We do not like to awaken
far from the fringe of the sea,
we who live upon small islands.

We like to rise up early,
quick in the agile mornings
and walk out only little distances
to look down at the water,

to know it is swaying near to us
with songs, and tides, and endless boatways,
and undulate[1] patterns and moods.

We want to be able to saunter beside it
slowpaced in burning sunlight,
barearmed, barefoot, bareheaded,

and to stoop down by the shallows
sifting the random water
between assaying[2] fingers
like farmers do with soil,

to think of turquoise mackerel
turning with consummate grace,
sleek and decorous
and elegant in high blue chambers.

---

[1] **undulate**   rising and falling
[2] **assaying**   testing, assessing

We want to be able to walk out into it,
to work in it,
dive and swim and play in it.

To row and sail
and pilot over its sandless highways,
and to hear
its call and murmurs wherever we may be.

All who have lived upon small islands
want to sleep and awaken
close to the fringe of the sea.

Further reading

Other poems by A. L. Hendriks can be found in *The Heinemann Book of Caribbean Poetry* (Heinemann, 1992). You may also like to read stories from the Caribbean. If so, you could do no better than *Summer Lightning and Other Stories* by Olive Senior (Longman, 1986).

For River

by Helen Hubbard

Helen Hubbard was 15 when she wrote this poem, which was influenced by the work of the poet Grace Nichols. Helen's poem was published in *The TES Book of Young Poets* (Times Supplements Ltd, 1999).

River could carry word
River could carry word

River babble in early morn
Word of mouth with every wave
River carrying her children
to her sea in firm embrace

But River don't seep her bank with mud
no River cover her back with blooms
from steaming jets of morning dew
and when cloud come to enclose moon
and darkness fall on her like blanket
River children go down to bed
Diamonds sparkle in fading sky

River flowing in her mind
to when she felt full gorged and high
From as a trickle, to a stream
how River roared in rampant open skies
River now at middle age wishes for those days
when her joy had not been dried
to that of old tempestuous tides
River wincing with eerie cries

and when sunbeams
wake her up with blinding light
River just creep and crawl
to a new day of thirsting love

but coming back to word
River could carry word
River could carry word
And we must follow River
to her word of mouth

Further reading

William Wordsworth's short poem *Composed Upon Westminster Bridge September 3, 1802* looks at a river from a different perspective from *For River*. As you read it you can imagine how the narrator is looking at his surroundings, as the Thames flows beneath him.

Sea Timeless Song
by Grace Nichols

Much of Grace Nichols' work is influenced by her early experiences
of the Caribbean and the natural world she knew. In this poem she
conjures up memories of the sea.

Hurricane come
and hurricane go
but sea . . . sea timeless
sea timeless
sea timeless
sea timeless
sea timeless

Hibiscus[1] bloom
then dry-wither so
but sea . . . sea timeless
sea timeless
sea timeless
sea timeless
sea timeless

Tourist come
and tourist go
but sea . . . sea timeless
sea timeless
sea timeless
sea timeless
sea timeless

Further reading

You may be familiar with Grace Nichols' work already, since many of
her poems can be found in anthologies. You could also read some of
her collections, such as *The Fat Black Woman's Poems* (Virago Press Ltd,
1984) and *Sunris* (Virago Press Ltd, 1996).

---

[1]**Hibiscus**   a shrub with brightly coloured flowers

The Highest Tide

by Jim Lynch

> Adolescent Miles feels a great affinity with the mudflats of Puget Sound, near where he lives. In an effort to cope with his parents' unhappy relationship and unable to sleep, he makes trips to the bay, searching for unusual marine specimens to sell to his local aquarium.

Near the oyster farm something happened that never failed to spook me in the dark. I saw a few dozen shore crabs scrambling near the rectangular, foot-high mesh fence around the judge's oyster beds. Crabs amused me in small crowds. It's when they clustered at night that they unhinged me, especially when they were in water where they moved twice as fast as on land. It was obvious there were more crabs – and bigger crabs – than usual, so I tried not to expand my range of vision too fast. It was no use. I saw hundreds, maybe thousands, assembling like tank battalions. I stepped back and felt their shells crunch beneath my feet and the wind pop out of me. Once I steadied, I flashed my headlamp on the oyster fence that three red rock crabs were aggressively scaling. It looked like a jail break with the biggest ringleaders leading the escape. I suddenly heard their clicking pincers clasping holds in the fence, jimmying their armored bodies higher. How had I missed that sound? The judge's oysters were under siege, but I couldn't bring myself to interfere. It felt like none of my business.

I picked my steps, knowing if I slipped and tumbled I'd feel them skittering around me as cool water filled my boots. I rounded the oyster beds, to the far side, relieved to find it relatively crab free. It was low tide by then, and I saw the water hesitating at its apex, neither leaving nor returning, patiently waiting for the gravitational gears to shift. Dozens of anxious clams started squirting in unison like they did whenever vibrating grains of sand warned them predators were approaching. I stopped and waited with them, to actually see the moment when the tide started returning with its invisible buffet of plankton for the clams, oysters, mussels and other filter feeders. It was right then, ankle deep in the Sound, feet numbing, eyes relaxed, that I saw the nudibranch.

In all my time on the flats I'd never seen one before. I'd read about them, sure. I'd handled them at aquariums but never in the wild, and I'd never even seen a photo of one this stunning.

It was just three inches long but with dozens of fluorescent, orange-tipped hornlike plumes jutting from the back of its see-through body that appeared to be lit from within.

Nudibranchs are often called the butterflies of the sea, but even that understates their dazzle. Almost everything else in the northern Pacific is dressed to blend with pale surroundings. Nudibranchs don't bother, in part because they taste so lousy they don't need camouflage to survive. But also, I decided right then, because their beauty is so startling it earns them a free pass, the same way everyday life brakes for peacocks, parade floats and supermodels.

I bagged that sea slug – it weighed nothing – and set it in my backpack next to the Jesus star. Then I gave the crabs a wide berth, found the moon snail, poked him in the belly until he contracted, bagged him and paddled south toward home beneath the almost-full moon.

And that's where it happened.

The dark mudflats loomed like wet, flattened dunes stretching deep into Skookumchuck Bay in front of our house. From a distance, they looked too barren to support sea life. Up close, they still did, unless you knew where to find the hearty clams, worms and tiny creatures that flourish in mud so fine that at least two Evergreen State College grads get stuck every June during their naked graduation prance across the bay's shallowest neck. I'm not sure why I decided to take a look. It was still an hour before sunrise, and I knew exactly what the bars looked like in the moonlight, but for some reason, I couldn't resist.

I heard it long before I saw it. It was an exhale, a release of sorts, and I instantly wondered if a whale was stranded again. We had a young minke stuck out there two summers prior, and it made similar noises until the tide rose high enough for rescuers to help free it. You would have thought the whole city had a baby, the pride people showed in guiding that little whale to deeper water. I looked for a hulking silhouette but couldn't find one. I waited, but there were no more sounds. Still, I went toward what I thought I'd heard, avoiding stepping into the mud until I had to. I knew the flats well enough to know I could get stuck just about anywhere. The general rule was you didn't venture out past the shells and gravel with an incoming tide. I sank up to my knees twice, and numbing water filled my boots.

South Sound is the warm end of the fjord because most of its bays are no deeper than forty feet and Skookumchuck is shallower still, but even in August the water rarely climbs much above fifty-five degrees and it can still take your breath. I kept stepping toward the one sound I'd heard, a growing part of me hoping I'd find nothing at all.

When I stopped to rest and yank up my socks, my headlamp crossed it. My first thought? A giant octopus.

Puget Sound has some of the biggest octopi in the world. They often balloon to a hundred pounds. Even the great Jacques Cousteau[1] himself came to study them. But when I saw the long tubular shape of its upper body and the tangle of tentacles below it, I knew it was more than an octopus. I came closer, within fifty feet, close enough to see its large cylindrical siphon quiver. I couldn't tell if it was making any sounds at that point because it was impossible to hear anything over the blood in my ears. My mother once told me that she had an oversized heart. I took her literally and assumed I was similarly designed because there were moments when mine sounded way too loud for a boy my size.

---

[1] **Jacques Cousteau**   Jacques-Yves Cousteau (1910–1997), a French filmmaker and ecologist

The creature's body came to a triangular point above narrow fins that lay flat on the mud like wings, but it was hard to be sure exactly where it all began or ended, or how long its tentacles truly were because I was afraid to pry my eyes off its jumble of arms for more than half a second. I didn't know whether I was within reach, and its arms were as big around as my ankle and lined with suckers the size of half dollars. If they even twitched I would have run. So, I was looking at it and not looking at it while my heart spangled my vision. I saw fragments, pieces, and tried to fuse them in my mind but couldn't be certain of the whole. I knew what it had to be, but I wouldn't allow myself to even think the two words. Then I gradually realized the dark shiny disc in the middle of the rubbery mass was too perfectly round to be mud or a reflection.

It was too late to smother my scream. Its eye was the size of a hubcap.

'It's a giant squid!'

Further reading

The Highest Tide (Bloomsbury Publishing Plc, 2008) has received a lot of publicity and is well worth reading. The author, Jim Lynch, has a website about his work: http://www.thehighesttide.com. He has also written the non-fiction *From May to September* (Wasteland Press, 2007), set in the 1950s, about a young man's transition from adolescence to adulthood.

Waterlog

by Roger Deakin

Waterlog: a Swimmer's Journey through Britain (Vintage, 2000) is, as its title suggests, an account of the author's first-hand experience of Britain's waterways in their many different forms. His book not only documents this unusual journey, but also captures the unique emotional experience of the swimmer and his proximity to nature.

The warm rain tumbled from the gutter in one of those midsummer downpours as I hastened across the lawn behind my house in Suffolk and took shelter in the moat. Breaststroking up and down the thirty yards of clear, green water, I nosed along, eyes just at water level. The frog's-eye view of rain on the moat was magnificent. Rain calms water, it freshens it, sinks all the floating pollen, dead bumblebees and other flotsam. Each raindrop exploded in a momentary, bouncing fountain that turned into a bubble and burst. The best moments were when the storm intensified, drowning birdsong, and a haze rose off the water as though the moat itself were rising to meet the lowering sky. Then the rain eased and the reflected heavens were full of tiny dancers: water sprites springing up on tiptoe like bright pins over the surface. It was raining water sprites.

It was at the height of this drenching in the summer of 1996 that the notion of a long swim through Britain began to form itself. I wanted to follow the rain on its meanderings about our land to rejoin the sea, to break out of the frustration of a lifetime doing lengths, of endlessly turning back on myself like a tiger pacing its cage.

Like the endless cycle of the rain, I would begin and end the journey in my moat, setting out in spring and swimming through the year. I would keep a log of impressions and events as I went.

My earliest memory of serious swimming is of being woken very early on holiday mornings with my grandparents in Kenilworth by a sudden rain of pebbles at my bedroom window aimed by my Uncle Laddie, who was a local swimming champion and had his own key to the outdoor pool. My cousins and I were reared on mythic tales of his exploits – in races, on high boards, or swimming far out to sea – so it felt an honour to swim with him. Long before the lifeguards arrived, we would unlock the wooden gate and set the straight, black, refracted lines on the bottom of the green pool snaking and shimmying. It was usually icy, but the magic of being first in is what I remember. 'We had the place to ourselves,' we would say with satisfaction afterwards over breakfast. Our communion with the water was all the more delightful for being free of charge. It was my first taste of unofficial swimming.

Years later, driven mad by the heat one sultry summer night, a party of us clambered over the low fence of the old open-air pool at Diss in Norfolk. We joined other silent, informal swimmers who had somehow stolen in, hurdling the dormant turnstiles, and now loomed past us in the water only to disappear again into the darkness like characters from *Under Milk Wood*. Such indelible swims are like dreams, and have the same profound effect on the mind and spirit. In the night sea at Walberswick I have seen bodies fiery with phosphorescent plankton striking through the neon waves like dragons.

The more I thought about it, the more obsessed I became with the idea of a swimming journey. I started to dream ever more exclusively of water. Swimming and dreaming were becoming indistinguishable. I grew convinced that following water, flowing with it, would be a way of getting under the skin of things, of learning something new. I might learn about myself, too. In water, all possibilities seemed infinitely extended.

When you swim, you feel your body for what it mostly is – water – and it begins to move with the water around it. No wonder we feel such sympathy for beached whales; we are beached at birth ourselves. To swim is to experience how it was before you were born. Once in the water, you are immersed in an intensely private world as you were in the womb. These amniotic[1] waters are both utterly safe and yet terrifying, for at birth anything could go wrong, and you are assailed by all kinds of unknown forces over which you have no control. This may account for the anxieties every swimmer experiences from time to time in deep water. A swallow dive off the high board into the void is an image that brings together all the contradictions of birth. The swimmer experiences the terror and the bliss of being born.

So swimming is a rite of passage, a crossing of boundaries: the line of the shore, the bank of the river, the edge of the pool, the surface itself. When you enter the water, something like metamorphosis happens. Leaving behind the land, you go through the looking-glass surface and enter a new world, in which survival, not ambition or desire, is the dominant aim. The lifeguards at the pool or the beach remind you of the thin line between waving and drowning. You see and experience things when you're swimming in a way that is completely different from any other. You are *in* nature, part and parcel of it, in a far more complete and intense way than on dry land, and your sense of the present is overwhelming. In wild water you are on equal terms with the animal world around you: in every sense,

---

[1]**amniotic**   the fluid that surrounds a foetus in the womb

on the same level. As a swimmer, I can go right up to a frog in the water and it will show more curiosity than fear. The damselflies and dragonflies that crowd the surface of the moat pointedly ignore me, just taking off for a moment to allow me to go by, then landing again in my wake.

Natural water has always held the magical power to cure. Somehow or other, it transmits its own self-regenerating powers to the swimmer. I can dive in with a long face and what feels like a terminal case of depression, and come out a whistling idiot. There is a feeling of absolute freedom and wildness that comes with the sheer liberation of nakedness as well as weightlessness in natural water, and it leads to a deep bond with the bathing-place.

Most of us live in a world where more and more places and things are signposted, labelled, and officially 'interpreted'. There is something about all this that is turning the reality of things into virtual reality. It is the reason why walking, cycling and swimming will always be subversive[2] activities. They allow us to regain a sense of what is old and wild in these islands, by getting off the beaten track and breaking free of the official version of things. A swimming journey would give me access to that part of our world which, like darkness, mist, woods or high mountains, still retains most mystery. It would afford me a different perspective on the rest of landlocked humanity.

My moat, where the journey first suggested itself, and really began, is fed by a vigorous spring eleven feet down, and purified by an entirely natural filtration system far superior to even the most advanced of swimming-pool technology. It is sustained by the plant and animal life you will find in any unpolluted fresh-water pond left to its own devices and given plenty of sunlight. There seems to have been a period, from the later Middle Ages until the seventeenth century,

---

[2]**subversive**   undermining, overthrowing

when moats became as fashionable in Suffolk as private pools are today. There are over thirty of them within a four-mile radius of the church in the nearby village of Cotton. Moats are now considered by historians like Oliver Rackham to have functioned as much as status symbols as anything else for the yeoman farmers who dug them. Mine was probably excavated when the house was built in the sixteenth century, and runs along the front and back of the house but not the sides. It had no defensive function except as a stock barrier. It would have yielded useful clay for building and formed a substantial reservoir, but it was certainly never intended for swimming. Its banks plunge straight down and it has no shallow end. At one end, where you climb in or out by a submerged wooden cart-ladder I have staked to the bank, a big willow presides, its pale fibrous roots waving in the water like sea anemones.

The moat is where I have bathed for years, swimming breaststroke for preference. I am no champion, just a competent swimmer with a fair amount of stamina. Part of my intention in setting out on the journey was not to perform any spectacular feats, but to try and learn something of the mystery D. H. Lawrence noticed in his poem 'The Third Thing':

Water is H_2O, hydrogen two parts, oxygen one,
But there is also a third thing, that makes it water
And nobody knows what that is.

For the best part of a year, the water would become my natural habitat. Otters sometimes set off across country in search of new territory, fresh water, covering as much as twelve miles in a night. I suppose there is part of all of us that envies the otter, the dolphin and the whale, our mammal cousins who are so much better adapted to water than we are, and seem to get so much more enjoyment from life than we do. If I could learn even a fraction of whatever they know, the journey would be richly repaid.

Packing, the night before I left, I felt something of the same apprehension and exhilaration as I imagine the otter might feel about going off into the blue. But, as with Ned Merrill in 'The Swimmer', my impulse to set off was simple enough at heart: 'The day was beautiful and it seemed to him that a long swim might enlarge and celebrate its beauty.'

Further reading

If you enjoyed this extract you might like to read the rest of the book. It was very well reviewed in the mainstream press and natural history journals.

Activities

Gone to Sea

Before you read

1   What is the sea like from a distance? Have you seen it from a plane, a cliff or a coastal road? What impression do you get of it? How does it inspire your imagination? Work with a partner and share your experiences and ideas about the sea or ocean.

What's it about?

2   What was William like as a baby? What was he like as a child? Who treated him well? Share your ideas with a partner.

3   Who encouraged William to swim? Why did the seals become important to him? With a partner, think of more than one reason.

4   William's leaving note says 'Gone to sea, where I belong'. Discuss with a partner the different meanings that his family could take from his message.

Thinking about the text

5   William did not feel good about himself because of the way he was treated. Who treated William badly? Who do you think caused him the greatest unhappiness? His father? Other children? Carry out a discussion in a small group to answer these questions. Agree and disagree and share opinions. Appoint someone to make notes. Then try to come to an agreement as a group, and report back to the class.

6   Think of several different ways people could have shown William respect and share your ideas with a partner. Then together draw up an advice sheet, entitled 'How to treat people well'.

Daughter of the Sea

Before you read

1   What stories, novels or films do you know that involve the sea? How important was the sea? How did it affect the characters? Share your ideas with others in a group.

What's it about?

Read the text and answer questions 2 to 4 with a partner. Make short notes.

2   Re-read the Prologue. Where is the story set? What do the people do, the women as well as the men?

3   What do you think Munroe has brought in his basket? Try to guess, sharing each other's ideas. You may think you know – but think again!

4   What part might Eilean play in future events? Use your imagination and think of three possibilities.

Thinking about the text

5   The language of the text creates atmosphere and mood. This is partly through the use of powerful verbs such as 'flung' and 'haunt' (page 69) and powerful adjectives such as 'jagged' and 'lashing' (page 69). Find three more of each, then write three sentences of your own about the seashore on a wild night. Use strong vocabulary to conjure up a vivid picture.

6   Write Jannet's diary entry (using the first person 'I') about the events in the chapter entitled 'The Freak Storm' (pages 70–72). Think about:
   ● the wild, stormy night
   ● the strange feelings she experiences as she waits for her husband
   ● the presence of Eilean o da Freya.

7   **a**   Design a backdrop for a play of *Daughter of the Sea*. Think about:
      ● what part of the text you would focus on
      ● what colours and shapes to include (for example, if you chose the shore, how would you convey the storm?)
   b   Write a caption for your backdrop. Describe the materials you used and what you would like the viewer to feel.

The Fringe of the Sea

Before you read

1   What is the sea like in different seasons? Work in a small group. Each member of the group should describe a different season. As a group, think of three adjectives to sum up each season.

What's it about?

2   In what country could the poem be set? List several that fit the details of the poem.

3   Re-read the last verse. What are the narrator's feelings? Why do you think he feels this way? Find more than one reason.

Thinking about the text

4   a   Study the image in verse five. What senses does it appeal to? What does it suggest to you about the farmer's relationship with the land? Think about the process of working the land as well as its natural qualities. What does the comparison suggest about the narrator's feelings about the sea? Discuss your ideas with a partner.

b   Write a role play in which a farmer meets the narrator of the poem and ask each other questions. Take a role each and perform your role play for another pair.

5   In verse one the reader is introduced to the main focus of the poem. The last verse draws the poem to a close. In what ways are the two verses similar? Make notes, thinking about the form (shape and patterns) as well as the meaning of the words.

6   Study the images in verse six. How is the beauty of nature described here? What do you think the 'high blue chambers' might be? Imagine you are a diver. Write a poem about what you see in the underwater world. Think about:

- colours
- shapes
- senses, particularly sight and touch.

Think about the form of your poem, such as how you could order the lines and verses, and how to open and close your poem.

For River

Before you read

1 Poets have often used rivers to evoke images of distance, the passing of time and change. Why do you think this is? Write down your views in a few sentences.

What's it about?

Read the text, then work with a partner to answer questions 2 to 4.

2 When we read the poem we feel as if the river can speak. What words suggest this?

3 What times of day are mentioned in the poem? What words suggest the passing of time?

4 There are many words in the poem that remind us of either the sound or the movement of water, or sometimes both. How many can you find? Draw up a list.

Thinking about the text

5 In the poem different techniques are used to create the feeling that water is moving. For example, the repetition of lines such as 'River could carry word' emphasises rhythm and pace. Look for other techniques that do the same, thinking carefully about the length of the lines and the punctuation, and the language used. Make notes, starting with 'Repetition' and the example.

6 There is a very strong voice in the poem and the effect of this is to give the river a distinctive character. Identify as many features as you can that help to do this. Think about:
- the type of English used
- what gender the river is
- how the river thinks and feels
- the sounds and images created (e.g. 'River babble').

7 Practise a performance of the poem with a partner. Decide:
- who will read what (e.g. a verse each)
- where you will slow down and where you will read more quickly
- how and where you will stand to deliver the poem.

Present your performance to another pair.

Sea Timeless Song

Before you read

1   Work with a partner. Describe to each other the different sounds the sea makes.

What's it about?

Read the text, then work with a partner to make short notes in answer to questions 2 to 4.

2   In what way does the title suit the poem's meaning?

3   What is the mood of the sea in the poem? How did you decide?

4   In what way could the poem tell us something about our own lives?

Thinking about the text

5   What comes and goes in the poem, in contrast to the sea? What other things in the natural world (including creatures) could contrast with the sea in a similar way? Jot down some examples. What has permanence like the sea, or could last for a very long time? Again jot down some examples.

6   Poetry often contains alliteration (words that begin with the same sound to create an effect, such as 'bees and butterflies').
   a   Alliteration is very important in this poem. Which is the clearest example? What aspects of the sea does it remind you of? Now find another example of alliteration in the poem.
   b   Create your own images of the sea using different sounds as alliteration.

7   Repetition helps to give the poem form. Look at the verse pattern. In what way is each verse similar? Repetition also suits the poem because it reminds us of the ebb and flow of the sea. Write a paragraph about the different kinds of repetition in the poem and how this links to the poem's subject.

8   Write your own poem about something in the natural world (it need not be the sea) using some of the techniques you have studied in questions 5 to 7: contrasting features, alliteration and repetition.

The Highest Tide

Before you read

1   What are mudflats? Why can they be dangerous and when? Why do you have to know them well before you explore them? Do you know of any treacherous mudflats that have been in the news? Discuss what you know in a small group.

What's it about?

Read the text, then work with a partner to make short notes in answer to questions 2 to 4.

2   What makes the mudflats particularly eerie when Miles is there?

3   What different kinds of creatures are found on the mudflats? Name a few of them. Which are the filter feeders? How do they feed?

4   What extraordinary creature does Miles find? What does it look like?

Thinking about the text

5   The mudflats seem like another world full of strange happenings.
   a   When the narrator describes the shore crabs' attempts to reach the oyster beds, what does he compare them to? What comical images are conjured up?
   b   Nudibranches are referred to as 'the butterflies of the sea'. Think of another imaginative description for them from the details in the text. In the same way, create names for:

   a clam   a moon snail   an octopus

6   Imagine you could shoot a documentary film of the bay.
   a   Identify particularly vivid sections of the text.
   b   Then select several images from the sections, giving the images titles (include Miles' discovery amongst them).
   c   Add notes justifying your choices.

7   You are one of the newspaper reporters that subsequently arrives on the scene. Write your report of Miles' discovery. Include coverage of the bay, a detailed description of the creature and its size, and Miles' reaction to events.

Waterlog

Before you read

1 Do you enjoy swimming? Would you describe yourself as a water baby or do you have a fear of water? Share your ideas in a group.

What's it about?

2 What experience gave Roger Deakin the idea for his journey?

3 Discuss with a partner what the writer means when he says that:
- we live in a world that is increasingly 'signposted, labelled and officially "interpreted"' (page 87)
- 'walking, cycling and swimming will always be subversive activities' (page 87).

4 Re-read the extract from the poem *The Third Thing* (page 88). Decide with your partner what the third thing about water could be.

Thinking about the text

5 The text opens with a description of rain in a midsummer downpour. Write a series of shape poems about rain. Choose words to capture:
- the way rain falls, in wind, for example
- how rain settles as pools and puddles
- how it forms in clouds or as mist.

6 a The writer uses expressive language skilfully. What do the following metaphors or similes refer to?
- It was raining water sprites (page 84)
- striking through the neon waves like dragons (page 85)
- waving in the water like sea anemones (page 88)
- going off into the blue (page 89).

   b Create your own images for the following:

   a moat   a dragonfly   an otter in water

7 Re-read the paragraph beginning 'My earliest memory of serious swimming' (page 85). Then write an essay that includes:
- a description of your first experiences of learning to swim or being in water, and how you felt (if you can't remember, imagine the experience)
- your thoughts on how swimming could benefit you emotionally as well as physically.

Compare and contrast

1 a We do not know whether *For River* was influenced by *Sea Timeless Song,* but Helen Hubbard mentions that Grace Nichols' poetry influenced her work. With a partner, discuss the similarities and differences between the two poems. Think about:
   ● what the poems are about
   ● rhythm and sound
   ● images and language.

   b Now imagine Helen Hubbard is being interviewed on radio about her poem. One of you should play the interviewer and one the poet. Try to imagine the scene in a radio studio. The interviewer should adopt a polite and self-assured tone. How might the poet feel? Confident? Nervous? Act out the interview.

2 The sea figures in a compelling way in many of the texts in this section. Which two texts do you think present the sea most powerfully? Write an essay comparing and contrasting the two texts, referring to or quoting from them as you need to.

3 Work with a small group. Each of you should choose a different text that evokes the mood and mystery of water and make brief notes about how the writer does this. For example, you could consider the rhythm of a poem or the language used to describe water. Then make a case for your chosen text as the best one and present it to the rest of your group. The group should then decide who has made the most convincing case and why.

4 Which of the fiction texts in *Water worlds* do you think is the most dramatic. Why? Choose a scene from one and turn it into a short play script. You will need to think about:
   ● which would be easiest to stage
   ● which text lends itself best to dialogue
   ● what to leave out and what to include
   ● what to include in the stage directions.

   Add one more character if you need to.

5 'Selkies' or 'silkies' are found in many Celtic folk tales. Find out what they are by using an encyclopaedia or the Internet. Then work in a group to discuss in what ways *Gone to Sea* and *Daughter of the Sea* are.

- similar to these tales
- different from these tales.

Appoint a member of your group to draw up a list of points under the headings 'Similar to' and 'Different from'.

3 Nature's power

Have you ever felt powerless in the face of nature – perhaps in a violent storm, on a freezing day or during a heat wave? In this section the texts cover aspects of nature's immensity and force, from hurricanes and high winds to Alaskan ice and the eruption of Vesuvius, detailed in Pliny's letter written nearly 2000 years ago.

Natural disasters have affected humanity since time began and we still have a fascination with them. But the texts also explore the sheer wonder of nature and its emotional impact on us; one text in particular, *Nature Cure*, highlights how nature can heal.

As you read these texts, think about the effects of nature's power in different parts of the world, and what qualities the main characters or people need to survive – but also how nature can support us through difficult times.

Activities

1   There is a fable by Aesop in which the sun and the wind argue about which is the more powerful. To decide the dispute they agree that whichever one can make a traveller remove his cloak first is the stronger. The wind's force makes the traveller pull his cloak more tightly around him, but the sun's heat makes the traveller remove it. So the moral of the fable is that persuasion is better than force. Write a modern version of the tale with the same moral. Use the power of nature in some way.

2   Active volcanoes can be found on various continents. Search for a map showing them; for example, go to http://www.geo.mtu.edu/volcanoes/world.html. Carry out further research into newspaper reports on one of the volcanoes shown.

3   Many people find that nature is an aid in times of sadness or stress. For example, quiet woods or city parks may provide comfort, wide open landscapes may broaden our perspective on life, and taking exercise in the countryside can reduce anxiety. What aspect of nature would help you in such circumstances and why? Discuss your thoughts in a small group.

Dust Storm

by Karen Hesse

The poem *Dust Storm* is part of a much longer book written in free verse, set in the 1930s. It tells the story of Billie Jo and her family, the calamities they face and their struggle to survive.

I never would have gone to see the show
if I had known a storm like this would come.
I didn't know when going in,
but coming out
a darker night I'd never seen.
I bumped into a box beside the Palace door
and scraped my shins,
then tripped on something in my path,
I don't know what,
and walked into a phone pole,
bruised my cheek.

The first car that I met was sideways in the road.
Bowed down, my eyes near shut,
trying to keep the dust out,
I saw his headlights just before I reached them.

The driver called me over and I felt my way,
following his voice.
He asked me how I kept the road.
'I feel it with my feet,' I shouted over the
roaring wind,
'I walk along the edge.
One foot on the road, one on the shoulder.'
And desperate to get home,
he straightened out his car,
and straddled tires on the road and off,
and slowly pulled away.

I kept along. I know that there were others
on the road,
from time to time I'd hear someone cry out,
their voices rose like ghosts on the howling wind;
no one could see. I stopped at neighbors'
just to catch my breath
and made my way from town
out to our farm.
Everyone said to stay
but I guessed
my father would
come out to find me
if I didn't show,
and get himself lost in the
raging dust and maybe die
and I
didn't want that burden on my soul.

Brown earth rained down
from sky.
I could not catch my breath
the way the dust pressed on my chest
and wouldn't stop.
The dirt blew down so thick
it scratched my eyes
and stung my tender skin,
it plugged my nose and filled inside my mouth.
No matter how I pressed my lips together,
the dust made muddy tracks
across my tongue.

But I kept on,
spitting out mud,
covering my mouth,
clamping my nose,
the dust stinging the raw and open
stripes of scarring on my hands,
and after some three hours I made it home.

Inside I found my father's note
that said he'd gone to find me
and if I should get home, to just stay put.
I hollered out the front door
and the back;
he didn't hear,
I didn't think he would.
The wind took my voice and busted it
into a thousand pieces,
so small
the sound
blew out over Ma and Franklin's grave,
thinner than a sigh.

I waited for my father through the night, coughing up
dust,
cleaning dust out of my ears,
rinsing my mouth, blowing mud out of my nose.

Joe De La Flor stopped by around four to tell me
they found one boy tangled in a barbed-wire fence,
another smothered in a drift of dust.

After Joe left I thought of the famous Lindberghs,[1]
and how their baby was killed and never came back
to them.
I wondered if my father would come back.

He blew in around six A.M.
It hurt,
the sight of him
brown with dirt,
his eyes as red as raw meat,
his feet bruised from walking in worn shoes
stepping where he couldn't see
on things that bit and cut into his flesh.

I tried to scare up[2] something we could eat,
but couldn't keep the table clear of dust.
Everything I set
down for our breakfast
was covered before we took a bite,
and so we chewed the grit and swallowed
and I thought of the cattle
dead from mud in their lungs,

---

[1] **Lindberghs**   Charles Lindbergh, a famous American aviator, and his wife
Anne Morrow Lindbergh; the kidnapping and death of their baby
Charles in the 1930s became a famous case
[2] **scare up**   use whatever is at hand to put something together

and I thought of the tractor
buried up to the steering wheel,
and Pete Guymon,
and I couldn't even recognize the man
sitting across from me,
sagging in his chair,
his red hair gray and stiff with dust,
his face deep lines of dust,
his teeth streaked brown with dust.
I turned the plates and glasses upside down,
crawled into bed, and slept.

Further reading

Why not read the other poems that form the story in *Out of the Dust* (Frances Lincoln Children's Books, 2007)? It absorbs the reader in the same way that a good novel does and is certainly recommended.

The famous *Let Us Now Praise Famous Men* (Penguin Classics, 2006), by James Agee and Walker Evans, also explores the lives of families living in the American 'dust bowl'. It provides a remarkable photographic and written record of their lives. One of Walker Evans' photographs (of Lucille Burroughs) appears on the cover of Karen Hesse's *Out of the Dust*.

Hurricane Gold
by Charlie Higson

Hurricane Gold is a *Young Bond* novel written by Charlie Higson. The main character is a youthful James, based on the adult James Bond devised by Ian Fleming. Here James finds himself at the house of the wealthy Jack Stone, in the town of Tres Hermanas on the Caribbean island of Lagrimas Negras. A hurricane is brewing. At the house are Jack's son JJ and his daughter Precious, when intruders break in.

In the belly of the storm

James watched as the young man waved his gun at Precious and JJ.

'Where's your father?' he yelled. 'Tell me or I'll hurt you.'

'He's not here,' wailed Precious. 'He's flown down south. He won't be back until after the storm.'

As Precious said the word 'storm' three things happened at once. There was a terrific crack of thunder, the whole house shook and the lights went out.

The storm had finally arrived.

Precious screamed. The young man snarled at her to shut up. There was just enough light coming through the window for James to see him grab the two children and drag them out of the room.

James stayed put, breathing heavily. The intruders seemed to have come prepared, but with luck they wouldn't know that he was here at all.

James waited in the Wendy house for a full five minutes. Once he was sure that the man wasn't coming back he crept out of his hiding place and tiptoed over to the playroom door.

He hardly needed to be quiet. The storm was making a fearsome racket as it buffeted the house. There was a cacophony[1] of different sounds; crashing, hissing, roaring, squealing, rumbling.

---

[1]**cacophony**   loud, harsh, clashing sounds

As he moved out into the corridor James felt the full force of the wind slam into the house like a physical object. He could actually feel the floor moving beneath his feet, and the walls seemed to sway and shudder. He glanced out of the window, but all he could see was a swirling maelstrom[2] of cloud and rain. There was a startling flash and another blast of thunder, then a gust of wind so powerful it blew the windows in. The rain followed, hosing down the corridor in horizontal bars. The walls were instantly soaked and a picture flew off the wall.

The noise from outside was like nothing the James had ever heard before, like boulders crashing down a mountainside. The wind was whipping around in the corridor and the house was vibrating as if at any moment it might crack up and be blown away.

---

[2]**maelstrom**   whirlpool

James dropped to his knees and crawled along the sodden carpet as bits of debris were hurled past his head.

He reached the stairs and slid down them on his backside in the darkness. He made it safely to the lower landing and peered out between the banisters into the hallway below.

The servants were being rounded up and herded into the dining room by two of the men. The raid had been planned like a military operation.

James was the only person who might be able to get out and go for help.

He backed away from the banisters, ducked into his bedroom and pulled the door shut.

He stood there for a moment, with his back to the door, breathing deeply. Rain was pouring in through the open window, and the carpet was soaked. There was already a large pool of water forming in the middle of the floor. James was sweating again. But it was a cold sweat, caused by fear, not heat. The temperature had dropped dramatically.

He considered his options and found that he had only one: to climb out and make a run for it.

He remembered seeing a little ornamental balcony outside and some thick jungly creeper up the side of the house.

He stepped towards the window, then suddenly threw himself to the floor as a piece of wood the size of a tabletop exploded through the window, spraying the room with jagged splinters. It was a broken door. The wind must have ripped it off another house and tossed it up here.

Over the sound of the storm James heard shouts and someone running up the stairs.

He quickly pulled the bedclothes off the bed and covered himself with them, leaving just enough space to see out.

He saw the door open and a pair of legs come in.

'It's just the storm,' the person shouted. 'Window's smashed. It's getting pretty hairy out there.'

The legs departed and the door slammed shut.

James crawled out from under the bedclothes. The room was strewn with bits of wood and shredded leaves. He battled his way to the window and looked out into the belly of the storm.

It was hopeless. He wouldn't last five minutes out there. Even if he made it out of the garden, which was unlikely, he doubted that there would be anybody who would be able to help him. No one would risk leaving the safety of their home to brave this storm.

The wind was throwing stuff in every direction. The palm trees were bent over, and, as he watched, a large shrub was uprooted and sent spinning across the lawn. It ended up tangled in the iron gates. Another, even stronger gust tore the gates loose from their hinges. They tumbled into the road and bounced off out of view.

Occasionally the wind would change direction, the clouds would break and he would get a glimpse of Tres Hermanas. No lights were showing. The buildings were a black tumble. The electricity must be out everywhere.

James couldn't tear his eyes away. He was mesmerized by the awesome power of the storm. A set of garden furniture rolled across the lawn and knocked over a statue. A large tree near the road, unable to bend, snapped in half and collapsed on to the perimeter wall, flattening it. All the tiles from the roof of a nearby outbuilding were plucked off and James only just managed to duck down out of the way before they came clattering against the side of the house as if thrown by some bad-tempered giant.

His face was wet and his eyes stung. He couldn't tell whether it was him or the house that was shaking.

Probably both.

He had never known a storm like this before.

Further reading

Charlie Higson has written several *Young Bond* novels, all of them fast-paced thrillers, with a likeable hero. *Hurricane Gold* is the fourth. If you read this book, you may well find you want to read the series. The first four are *Silver Fin* (2005), *Blood Fever* (2006) and *Double or Die* (2007).

To Build a Fire

by Jack London

In this short story by the American writer Jack London, a man accompanied by his husky dog is travelling through the Yukon in freezing weather.

Day had broken cold and gray, exceedingly cold and gray, when the man turned aside from the main Yukon trail and climbed the high earth-bank, where a dim and little-travelled trail led eastward through the fat spruce timberland. It was a steep bank, and he paused for breath at the top, excusing the act to himself by looking at his watch. It was nine o'clock. There was no sun nor hint of sun, though there was not a cloud in the sky. It was a clear day, and yet there seemed an intangible[1] pall over the face of things, a subtle gloom that made the day dark, and that was due to the absence of sun. This fact did not worry the man. He was used to the lack of sun. It had been days since he had seen the sun, and he knew that a few more days must pass before that cheerful orb, due south, would just peep above the sky-line and dip immediately from view.

The man flung a look back along the way he had come. The Yukon lay a mile wide and hidden under three feet of ice. On top of this ice were as many feet of snow. It was all pure white, rolling in gentle undulations where the ice-jams of the freeze-up had formed. North and south, as far as his eye could see, it was unbroken white, save for a dark hair-line that curved and twisted from around the spruce-covered island to the south, and that curved and twisted away into the north, where it disappeared behind another spruce-covered island. This dark hair-line was the trail – the main trail – that led south five hundred miles to the Chilcoot Pass, Dyea, and salt water; and that led north seventy miles to Dawson, and still on to the north a

---

[1]**intangible**   subtle, difficult to define

thousand miles to Nulato, and finally to St. Michael on Bering Sea, a thousand miles and half a thousand more.

But all this – the mysterious, far-reaching hair-line trail, the absence of sun from the sky, the tremendous cold, and the strangeness and weirdness of it all – made no impression on the man. It was not because he was long used to it. He was a new-comer in the land, a *chechaquo*, and this was his first winter. The trouble with him was that he was without imagination. He was quick and alert in the things of life, but only in the things, and not in the significances. Fifty degrees below zero meant eighty-odd degrees of frost. Such fact impressed him as being cold and uncomfortable, and that was all . . . Fifty degrees below zero stood for a bite of frost that hurt and that must be guarded against by the use of mittens, ear-flaps, warm moccasins and thick socks. Fifty degrees below zero was to him just precisely fifty degrees below zero. That there should be anything more to it than that was a thought that never entered his head.

As he turned to go on, he spat speculatively. There was a sharp explosive crackle that startled him. He spat again. And again, in the air, before it could fall to the snow, the spittle crackled. He knew that at fifty below spittle crackled on the snow, but this spittle had crackled in the air. Undoubtedly it was colder than fifty below – how much colder he did not know. But the temperature did not matter. He was bound for the old claim on the left fork of Henderson Creek, where the boys were already. They had come over across the divide from the Indian Creek country, while he had come the roundabout way to take a look at the possibilities of get-ting out logs in the spring from the islands in the Yukon. He would be in to camp by six o'clock; a bit after dark, it was true, but the boys would be there, a fire would be going, and a hot supper would be ready. As for lunch, he pressed his hand against the protruding bundle under his jacket. It was also under his shirt, wrapped up in a handkerchief and lying against the naked skin. It was the only way to keep the biscuits from freezing. He smiled agreeably to him-self as he thought of those biscuits, each cut open and sopped in bacon grease, and each enclosing a generous slice of fried bacon.

He plunged in among the big spruce trees. The trail was faint. A foot of snow had fallen since the last sled had passed over, and he was glad he was without a sled, travelling light. In fact, he carried nothing but the lunch wrapped in the handkerchief. He was surprised, however, at the cold. It certainly was cold, he concluded, as he rubbed his numb nose and cheek-bones with his mittened hand. He was a warm-whiskered man, but the hair on his face did not protect the high cheek-bones and the eager nose that thrust itself aggressively into the frosty air.

At the man's heels trotted a dog, a big native husky, the proper wolf-dog, gray-coated and without any visible or temperamental difference from its brother, the wild wolf. The animal was depressed by the tremendous cold. It knew that it was no time for travelling. Its instinct told it a truer tale than was told to the man by the man's judgement. In reality, it was not merely colder than fifty below zero; it was colder than sixty below, than seventy below. It was seventy-five below zero. Since the freezing point is thirty-two above zero, it meant that one hundred and seven degrees of frost obtained. The dog did not know anything about thermometers . . .

The frozen moisture of its breathing had settled on its fur in a fine powder of frost, and especially were its jowls, muzzle and eyelashes whitened by its crystalled breath. The man's red beard and mustache were likewise frosted, but more solidly, the deposit taking the form of ice and increasing with every warm, moist breath he exhaled. Also, the man was chewing tobacco, and the muzzle of ice held his lips so rigidly that he was unable to clear his chin when he expelled the juice. The result was that a crystal beard of the color and solidity of amber was increasing its length on his chin. If he fell down it would shatter itself, like glass, into brittle fragments. But he did not mind the appendage. It was the penalty all tobacco-chewers paid in that country, and he had been out before in two cold snaps. They had not been so cold as this, he knew, but by the spirit thermometer at Sixty Mile he knew they had been registered at fifty below and at fifty-five.

He held on through the level stretch of woods for several miles, crossed a wide flat of niggerheads, and dropped down a bank to the frozen bed of a small stream. This was Henderson Creek, and he knew he was ten miles from the forks. He looked at his watch. It was ten o'clock. He was making four miles an hour, and he calculated that he would arrive at the forks at half-past twelve. He decided to celebrate that event by eating his lunch there.

The dog dropped in again at his heels, with a tail drooping discouragement, as the man swung along the creek-bed. The furrow of the old sled-trail was plainly visible, but a dozen inches of snow covered the marks of the last runners. In a month no man had come up or down that silent creek. The man held steadily on. He was not much given to thinking, and just then particularly he had nothing to think about save that he would eat lunch at the forks and that at six o'clock he would be in camp with the boys. There was nobody to talk to; and, had there been, speech would have been impossible because of the ice-muzzle on his mouth. So he continued monotonously to chew tobacco and to increase the length of his amber beard.

Once in a while the thought reiterated itself that it was very cold and that he had never experienced such cold. As he walked along he rubbed his cheek-bones and nose with the back of his mittened hand. He did this automatically, now and again changing hands. But rub as he would, the instant he stopped his cheek-bones went numb, and the following instant the end of his nose went numb. He was sure to frost his cheeks; he knew that, and experienced a pang of regret that he had not devised a nose-strap of the sort Bud wore in cold snaps. Such a strap passed across the cheeks, as well, and saved them. But it didn't matter much, after all. What were frosted cheeks? A bit painful, that was all; they were never serious.

Empty as the man's mind was of thoughts, he was keenly observant, and he noticed the changes in the creek, the curves and bends and timber-jams, and always he sharply noted where he placed his feet. Once, coming around a bend, he shied abruptly, like a startled horse, curved away from the place where he had been walking, and retreated several paces back along the trail. The creek he knew was frozen clear to the bottom – no creek could contain water in that arctic winter – but he knew also that there were springs that bubbled out from the hillsides and ran along under the snow and on top of the ice of the creek. He knew that the coldest snaps never froze these springs, and he knew likewise their danger. They were traps. They hid pools of water under the snow that might be three inches deep, or three feet. Sometimes a skin of ice half an inch thick covered them, and in turn was covered by the snow. Sometimes there were alternate layers of water and ice-skin, so that when one broke through he kept on breaking through for a while, sometimes wetting himself to the waist.

That was why he had shied in such a panic. He had felt the give under his feet and heard the crackle of a snow-hidden ice-skin. And to get his feet wet in such a temperature meant trouble and danger. At the very least it meant delay, for he would be forced to stop and build a fire, and under its protection to bare

his feet while he dried his socks and moccasins. He stood and studied the creek-bed and its banks, and decided that the flow of water came from the right. He reflected awhile, rubbing his nose and cheeks, then skirted to the left, stepping gingerly and testing the footing for each step. Once clear of the danger, he took a fresh chew of tobacco and swung along at his four-mile gait.

In the course of the next two hours he came upon several similar traps. Usually the snow above the hidden pools had a sunken, candied appearance that advertised the danger. Once again, however, he had a close call; and once, suspecting danger, he compelled the dog to go in front. The dog did not want to go. It hung back until the man shoved it forward, and then it went quickly across the white, unbroken surface. Suddenly it broke through, floundered to one side, and got away to firmer footing. It had wet its fore-feet and legs, and almost immediately the water that clung to it turned to ice. It made quick efforts to lick the ice off its legs, then dropped down in the snow and began to bite out the ice that had formed between the toes. This was a matter of instinct. To permit the ice to remain would mean sore feet. It did not know this. It merely obeyed the mysterious prompting that arose from the deep crypts of its being. But the man knew, having achieved a judgement on the subject, and he removed the mitten from his right hand and helped to tear out the ice particles. He did not expose his fingers more than a minute, and was astonished at the swift numbness that smote them. It certainly was cold. He pulled on the mitten hastily, and beat the hand savagely across his chest.

At twelve o'clock the day was at its brightest. Yet the sun was too far south on its winter journey to clear the horizon. The bulge of the earth intervened between it and Henderson Creek, where the man walked under a clear sky at noon and cast no shadow. At half-past twelve, to the minute, he arrived at the forks of the creek. He was pleased at the speed he had made. If he kept it up, he would certainly be with the boys by six. He unbuttoned his jacket and shirt and drew forth his lunch. The action consumed no more than a quarter of a minute, yet in that brief moment the numbness laid hold of the exposed fingers. He did not put the mitten on, but, instead, struck the fingers a dozen sharp smashes against his leg. Then he sat down on a snow-covered log to eat. The sting that followed upon the striking of his fingers against his leg ceased so quickly that he was startled. He had had no chance to take a bite of biscuit. He struck the fingers repeatedly and returned them to the mitten, baring the other hand for the purpose of eating. He tried to take a mouthful, but the ice-muzzle prevented. He had forgotten to build a fire and thaw out. He chuckled at his foolishness, and as he chuckled he noted the numbness creeping into the exposed fingers. Also, he noted that the stinging which had first come to his toes when he sat down was already passing away. He wondered whether the toes were warm or numb. He moved them inside the moccasins and decided that they were numb.

He pulled the mitten on hurriedly and stood up. He was a bit frightened. He stamped up and down until the stinging returned into the feet. It certainly was cold, was his thought. That man from Sulphur Creek had spoken the truth when telling how cold it sometimes got in the country. And he had laughed at him at the time! That showed one must not be too sure of things. There was no mistake about it, it *was* cold. He strode up and down, stamping his feet and threshing his arms, until reassured by the returning warmth. Then he got out matches and proceeded to make a fire. From the undergrowth, where high water of the previous spring had lodged a supply of seasoned twigs, he got his fire-wood. Working carefully from a small beginning, he soon had a roaring fire, over which he thawed the ice from his face and in the protection of which he ate his biscuits. For the moment the cold of space was outwitted. The dog took satisfaction in the fire, stretching out close enough for warmth and far enough away to escape being singed.

When the man had finished, he filled his pipe and took his comfortable time over a smoke. Then he pulled on his mittens, settled the ear-flaps of his cap firmly about his ears, and took the creek trail up the left fork. The dog was disappointed and yearned back towards the fire. This man did not know cold. Possibly all the generations of his ancestry had been ignorant of cold, of real cold, of cold one hundred and seven degrees below freezing-point. But the dog knew; all its ancestry knew, and it had inherited the knowledge. And it knew that it was not good to walk abroad in such fearful cold. It was the time to lie snug in a hole in the snow and wait for a curtain of cloud to be drawn across the face of outer space whence this cold came . . .

The man took a chew of tobacco and proceeded to start a new amber beard. Also, his moist breath quickly powdered with white his mustache, eyebrows and lashes. There did not seem to be so many springs on the left fork of the Henderson, and for half an hour the man saw no signs of any. And then it happened. At a place where there were no signs, where the soft, unbroken snow seemed to advertise solidity beneath, the man broke through. It was not deep. He wet himself halfway to the knees before he floundered out to the firm crust.

He was angry, and cursed his luck aloud. He had hoped to get into camp with the boys at six o'clock, and this would delay him an hour, for he would have to build a fire and dry out his foot-gear. This was imperative at that low temperature – he knew that much; and he turned aside to the bank, which he climbed. On top, tangled in the underbrush about the trunks of several small spruce trees, was a high-water deposit of dry fire-wood – sticks and twigs, principally, but also larger portions of seasoned branches and fine, dry, last-year's grasses. He threw down several large pieces on top of the snow. This served for a foundation and prevented the young flame from drowning itself in the snow it otherwise would melt. The flame he got by touching a match to a small shred of birch-bark that he took from his pocket. This burned even more readily than paper. Placing it on the foundation, he fed the young flame with wisps of dry grass and with the tiniest dry twigs.

He worked slowly and carefully, keenly aware of his danger. Gradually, as the flame grew stronger, he increased the size of the twigs with which he fed it. He squatted in the snow, pulling the twigs out from their entanglement in the brush and feeding directly at the flame. He knew there must be no failure. When it is seventy-five below zero, a man must not fail in his first attempt to build a fire – that is, if his feet are wet. If his feet are dry, and he fails, he can run along the trail for half a mile and restore his circulation. But the circulation of wet and freezing feet cannot be restored by running when it is seventy-five below. No matter how fast he runs, the wet feet will freeze the harder.

All this the man knew. The old-timer on Sulphur Creek had told him about it the previous fall, and now he was appreciating the advice. Already all sensation had gone out of his feet. To build the fire he had been forced to remove his mittens, and the fingers had quickly gone numb. His pace of four miles an hour had kept his heart pumping blood to the surface of his body and to all the extremities. But the instant he stopped, the action of the pump eased down. The cold of space smote the unprotected tip of the planet, and he, being on that unprotected tip, received the full force of the blow. The blood of his body recoiled before it. The blood was alive, like the dog, and like the dog it wanted to hide away and cover itself up from the fearful cold. So long as he walked four miles an hour, he pumped that blood, willy-nilly, to the surface; but now it ebbed away and sank down into the recesses of his body. The extremities were the first to feel its absence. His wet feet froze the faster, and his exposed fingers numbed the faster, though they had not yet begun to freeze. Nose and cheeks were already freezing, while the skin of all his body chilled as it lost its blood.

But he was safe. Toes and nose and cheeks would be only touched by the frost, for the fire was beginning to burn with strength. He was feeding it with twigs the size of his finger. In another minute he would be able to feed it with branches the size of his wrist, and then he could remove his wet foot-gear, and, while it dried, he could keep his naked feet warm by the fire, rubbing them at first, of course, with snow. The fire was a success. He was safe. He remembered the advice of the old-timer on Sulphur Creek, and smiled. The old-timer had been very serious in laying down the law that no man must travel alone in the Klondike after fifty below. Well, here he was; he had had the accident; he was alone; and he had saved himself. Those old-timers were rather womanish, some of them, he thought. All a man had to do was to keep his head, and he was all right. Any man who was a man could travel alone. But it was surprising, the rapidity with which his cheeks and nose were freezing. And he had not thought his fingers could go lifeless in so short a time. Lifeless they were, for he could scarcely make them move together to grip a twig, and they seemed remote from his body and from him. When he touched a twig, he had to look and see whether or not he had hold of it. The wires were pretty well down between him and his finger-ends.

All of which counted for little. There was the fire, snapping and crackling and promising life with every dancing flame. He started to untie his moccasins. They were coated with ice; the thick German socks were like sheaths of iron halfway to the knees; and the moccasin strings were like rods of steel all twisted and knotted as by some conflagration. For a moment he tugged with his numb fingers, then, realizing the folly of it, he drew his sheath knife.

But before he could cut the strings, it happened. It was his own fault or, rather, his mistake. He should not have built the fire under the spruce tree. He should have built it in the open. But it had been easier to pull the twigs from the brush and drop them directly on the fire. Now the tree under which he had done this carried a weight of snow on its boughs. No wind had blown for weeks, and each bough was fully freighted. Each time he had pulled a twig he had communicated a slight agitation to the tree – an imperceptible agitation, so far as he was concerned, but an agitation sufficient to bring about the disaster. High up in the tree one bough capsized its load of snow. This fell on the boughs beneath, capsizing them. This process continued, spreading out and involving the whole tree. It grew like an avalanche, and it descended without warning upon the man and the fire, and the fire was blotted out! Where it had burned was a mantle of fresh and disordered snow.

The man was shocked. It was as though he had just heard his own sentence of death. For a moment he sat and stared at the spot where the fire had been. Then he grew very calm. Perhaps the old-timer on Sulphur Creek was right. If he had only had a trail-mate he would have been in no danger now. The trail-mate could have built the fire. Well, it was up to him to build the fire over again, and this second time there must be no failure. Even if he succeeded, he would most likely lose some toes. His feet must be badly frozen by now, and there would be some time before the second fire was ready.

Such were his thoughts, but he did not sit and think them. He was busy all the time they were passing through his mind. He made a new foundation for a fire, this time in the open, where no treacherous tree could blot it out. Next, he gathered dry grasses and tiny twigs from the high-water flotsam. He could not bring his fingers together to pull them out, but he was able to gather them by the handful. In this way he got many rotten twigs and bits of green moss that were undesirable, but it was the best he could do. He worked methodically, even collecting an armful of the larger branches to be used later when the fire gathered strength. And all the while the dog sat and watched him, a certain yearning wistfulness in its eyes, for it looked upon him as the fire-provider, and the fire was slow in coming.

When all was ready, the man reached in his pocket for a second piece of birch-bark. He knew the bark was there, and, though he could not feel it with his fingers, he could hear its crisp rustling as he fumbled for it. Try as he would, he could not clutch hold of it. And all the time, in his consciousness, was the knowledge that each instant his feet were freezing. This thought tended to put him in a panic, but he fought against it and kept calm. He pulled on his mittens with his teeth, and threshed his arms back and forth, beating his hands with all his might against his sides. He did this sitting down, and he stood up to do it; and all the while the dog sat in the snow, its wolf-brush of a tail curled around warmly over its forefront, its sharp wolf-ears pricked forward intently as it watched the man. And the man, as he beat and threshed with his arms and hands, felt a great surge of envy as he regarded the creature that was warm and secure in its natural covering.

After a time he was aware of the first faraway signals of sensation in his beaten fingers. The faint tingling grew stronger till it evolved into a stinging ache that was excruciating, but which the man hailed with satisfaction. He stripped the mitten from his right hand and fetched forth the birch-bark. The exposed fingers were quickly going numb again. Next he brought out his bunch of sulphur matches. But the tremendous cold had already driven the life out of his fingers. In his effort to separate one match from the others, the whole bunch fell in the snow. He tried to pick it out of the snow, but failed. The dead fingers could neither touch nor clutch. He was very careful. He drove the thought of his freezing feet, and nose, and cheeks, out of his mind, devoting his whole soul to the matches. He watched, using the sense of vision in place of that touch, and when he saw his fingers on each side the bunch, he closed them – that is, he willed to close them, for the wires were down, and the fingers did not obey. He pulled the mitten on the right hand, and beat it fiercely against his knee. Then with both mittened hands, he scooped the bunch of matches, along with much snow, into his lap. Yet he was no better off.

After some manipulation he managed to get the bunch between the heels of his mittened hands. In this fashion he carried it to his mouth. The ice crackled and snapped when by a violent effort he opened his mouth. He drew the lower jaw in, curled the upper lip out of the way, and scraped the bunch with his upper teeth in order to separate a match. He succeeded in getting one, which he dropped on his lap. He was no better off. He could not pick it up. Then he devised a way. He picked it up in his teeth and scratched it on his leg. Twenty times he scratched before he succeeded in lighting it. As it flamed he held it with his teeth to the birch bark. But the burning brimstone went up his nostrils and into his lungs, causing him to cough spasmodically. The match fell into the snow and went out.

The old-timer on Sulphur Creek was right, he thought in the moment of controlled despair that ensued: after fifty below, a man should travel with a partner. He beat his hands, but failed in exciting any sensation. Suddenly he bared both hands, removing the mittens with his teeth. He caught the whole bunch between the heels of his hands. His arm-muscles not being frozen enabled him to press the hand-heels tightly against the matches. Then he scratched the bunch along his leg. It flared into flame, seventy sulphur matches at once! There was no wind to blow them out. He kept his head to one side to escape the strangling fumes, and held the blazing bunch to the birch bark. As he so held it, he became aware of sensation in his hand. His flesh was burning. He could smell it. Deep down below the surface he could feel it. The sensation developed into pain that grew acute. And still he endured it, holding the flame of the matches clumsily to the bark that would not light readily because his own burning hands were in the way, absorbing most of the flame.

At last, when he could endure no more, he jerked his hands apart. The blazing matches fell sizzling into the snow, but the birch bark was alight. He began laying dry grasses and the tiniest twigs on the flame. He could not pick and choose, for he had to lift the fuel between the heels of his hands. Small pieces of rotten wood and green moss clung to the twigs, and he bit them off as well as he could with his teeth. He cherished the flame carefully and awkwardly. It meant life, and it must not perish. The withdrawal of blood from the surface of his body now made him begin to shiver, and he grew more awkward. A large piece of green moss fell squarely on the little fire. He tried to poke it out with his fingers, but his shivering frame made him poke too far, and he disrupted the nucleus of the little fire, the burning grasses and tiny twigs separating and scattering. He tried to poke them together again, but in spite of the tenseness of the effort, his shivering got away with him, and the twigs were hopelessly scattered. Each twig gushed a puff of smoke and went out. The fire provider had failed. As he looked apathetically about him, his eyes chanced on the dog, sitting across the ruins of the fire from him, in the snow, making restless, hunching movements, slightly lifting one forefoot and then the other, shifting its weight back and forth on them with wistful eagerness.

The sight of the dog put a wild idea into his head. He remembered the tale of the man, caught in a blizzard, who killed a steer and crawled inside the carcass, and so was saved. He would kill the dog and bury his hands in the warm body until the numbness went out of them. Then he could build another fire. He spoke to the dog, calling it to him; but in his voice was a strange note of fear that frightened the animal, who had never known the man to speak in such a way before. Something was the matter, and its suspicious nature sensed danger – it knew not what danger, but somewhere, somehow, in its brain arose an apprehension of the man. It flattened its ears down at the sound of the man's voice, and its restless, hunching movements and the liftings and shiftings of its fore-feet became more pronounced; but it would not come to the man. He got on his hands and knees and crawled towards the dog. This unusual posture again excited suspicion, and the animal sidled mincingly away.

The man sat up in the snow for a moment and struggled for calmness. Then he pulled on his mittens, by means of his teeth, and got upon his feet. He glanced down at first in order to assure himself that he was really standing up, for the absence of sensation in his feet left him unrelated to the earth. His erect position in itself started to drive the webs of suspicion from the dog's mind; and when he spoke peremptorily, with the sound of whip lashes in his voice, the dog rendered its customary allegiance and came to him. As it came within reaching distance, the man lost his control. His arms flashed out to the dog, and he experienced genuine surprise when he discovered that his hands could not clutch, that there was neither bend nor feeling in the fingers. He had forgotten for the moment that they were frozen and that they were freezing more and more. All this happened quickly, and before the animal could get away, he encircled its body with his arms. He sat down in the snow, and in this fashion held the dog, while it snarled and whined and struggled.

But it was all he could do, hold its body encircled in his arms and sit there. He realized he could not kill the dog. There was no way to do it. With his helpless hands he could neither draw nor hold his sheath-knife nor throttle the animal. He released it, and it plunged wildly away, with tail between its legs, and still snarling. It halted forty feet away and surveyed him curiously, with ears sharply pricked forward. The man looked down at his hands in order to locate them, and found them hanging on the ends of his arms. It struck him as curious that one should have to use his eyes in order to find out where his hands were. He began threshing his arms back and forth, beating the mittened hands against his sides. He did this for five minutes, violently, and his heart pumped enough blood up to the surface to put a stop to his shivering. But no sensation was aroused in the hands. He had an impression that they hung like weights on the ends of his arms, but when he tried to run the impression down, he could not find it.

A certain fear of death, dull and oppressive, came to him. This fear quickly became poignant as he realized that it was no longer a mere matter of freezing his fingers and toes, or of losing his hands and feet, but that it was a matter of life and death with the chances against him. This threw him into a panic, and he turned and ran up the creek-bed along the old, dim trail. The dog joined in behind him and kept up with him. He ran blindly, without intention, in fear such as he had never known in his life. Slowly, as he ploughed and floundered through the snow, he began to see things again – the banks of the creek, the old timber-jams, the leafless aspens, and the sky.

The running made him feel better. He did not shiver. Maybe, if he ran on, his feet would thaw out; and, anyway, if he ran far enough, he would reach camp and the boys. Without doubt he would lose some fingers and toes and some of his face; but the boys would take care of him, and save the rest of him when he got there. And at the same time there was another thought in his mind that said he would never get to the camp and the boys; that it was too many miles away, that the freezing had too great a start on him, and that he would soon be stiff and dead. This thought he kept in the background and refused to consider. Sometimes it pushed itself forward and demanded to be heard, but he thrust it back and strove to think of other things.

It struck him as curious that he could run at all on feet so frozen that he could not feel them when they struck the earth and took the weight of his body. He seemed to himself to skim along above the surface, and to have no connection with the earth. Somewhere he had once seen a winged Mercury, and he wondered if Mercury felt as he felt when skimming over the earth.

His theory of running until he reached camp and the boys had one flaw in it: he lacked the endurance. Several times he stumbled, and finally he tottered, crumpled up, and fell. When he tried to rise, he failed. He must sit and rest, he decided, and next time he would merely walk and keep on going. As he sat and regained his breath, he noted that he was feeling quite warm and comfortable. He was not shivering, and it even seemed that a warm glow had come to his chest and trunk. And yet, when he touched his nose or cheeks, there was no sensation. Running would not thaw them out. Nor would it thaw out his hands and feet. Then the thought came to him that the frozen portions of his body must be extending. He tried to keep this thought down, to forget it, to think of something else; he was aware of the panicky feeling that it caused, and he was afraid of the panic. But the thought asserted itself, and persisted, until it produced a vision of his body totally frozen. This was too much, and he made another wild run along the trail. Once he slowed down to a walk, but the thought of the freezing extending itself made him run again.

And all the time the dog ran with him, at his heels. When he fell down a second time, it curled its tail over its forefeet and sat in front of him, facing him, curiously eager and intent. The warmth and security of the animal angered him, and he cursed it till it flattened down its ears appeasingly. This time the shivering came more quickly upon the man. He was losing in his battle with the frost. It was creeping into his body from all sides. The thought of it drove him on, but he ran no more than a hundred feet, when he staggered and pitched headlong. It was his last panic. When he had recovered his breath and control, he sat up and entertained in his mind the conception of meeting death with dignity. However, the conception did not come to him in such terms. His idea of it was that he had been making a fool of himself, running around like a chicken with its head cut off – such was the simile that occurred to him. Well, he was bound to freeze anyway, and he might as well take it decently. With this new-found peace of mind came the first glimmerings of drowsiness. A good idea, he thought, to sleep off to death. It was like taking an anaesthetic. Freezing was not so bad as people thought. There were lots worse ways to die.

He pictured the boys finding his body next day. Suddenly he found himself with them – coming along the trail looking for himself. And, still with them, he came around a turn in the trail and found himself lying in the snow. He did not belong with himself any more, for even then he was out of himself, standing with the boys and looking at himself in the snow. It certainly was cold, was his thought. When he got back to the States he could tell the folks what real cold was. He drifted on from this to a vision of the old-timer on Sulphur Creek. He could see him quite clearly, warm and comfortable, and smoking a pipe.

'You were right, old hoss; you were right,' the man mumbled to the old-timer of Sulphur Creek.

Then the man drowsed off into what seemed to him the most comfortable and satisfying sleep he had ever known. The dog sat facing him and waiting. The brief day drew to a close in a long, slow twilight. There were no signs of a fire to be made, and, besides, never in the dog's experience had it known a man to sit like that in the snow and make no fire. As the twilight drew on, its eager yearning for the fire mastered it, and with a great lifting and shifting of forefeet, it whined softly, then flattened its ears down in anticipation of being chidden by the man. But the man remained silent. Later the dog whined loudly. And still later it crept close to the man and caught the scent of death. This made the animal bristle and back away. A little longer it delayed, howling under the stars that leaped and danced and shone brightly in the cold sky. Then it turned and trotted up the trail in the direction of the camp it knew, where were the other food-providers and fire-providers.

Further reading

If you enjoyed the setting of this short story and the depiction of the husky dog, you might like to read Jack London's novel, *White Fang* (Puffin Classics, 2008), which tells the story of a wild wolf dog. A film of the same name and based on the book was made in 1991, starring Ethan Hawke.

Wind

by Ted Hughes

Ted Hughes was a famous English poet. Much of his poetry is rooted in nature and its harsh beauty, particularly the landscape and the creatures of his native Yorkshire.

This house has been far out at sea all night,
The woods crashing through darkness, the booming hills,
Winds stampeding the fields under the window
Floundering black astride and blinding wet

Till day rose; then under an orange sky
The hills had new places, and wind wielded
Blade-light, luminous black and emerald,
Flexing like the lens of a mad eye.

At noon I scaled along the house-side as far as
The coal-house door. Once I looked up –
Through the brunt wind that dented the balls of my eyes
The tent of the hills drummed and strained its guyrope,

The fields quivering, the skyline a grimace,
At any second to bang and vanish with a flap;
The wind flung a magpie away and a black-
Back gull bent like an iron bar slowly. The house

Rang like some fine green goblet in the note
That any second would shatter it. Now deep
In chairs, in front of the great fire, we grip
Our hearts and cannot entertain book, thought,

Or each other. We watch the fire blazing,
And feel the roots of the house move, but sit on,
Seeing the window tremble to come in,
Hearing the stones cry out under the horizons.

Further reading

Ted Hughes has written fine, lasting poems about nature. In his *Collected Poems* (Faber and Faber, 2005) you can find *Hawk Roosting*, *Crow*, *The Thought-Fox*, *An Otter* and many more.

The Hail Storm in June 1831

by John Clare

John Clare was born into a poor household; he was the son of a farm labourer and he himself worked on the land from childhood. He suffered poor physical and mental health in later life. The natural world had a profound effect on him and he always had to be near it to find comfort. He is considered to be one of the most important poets of the 19th century and of nature. In this sonnet he describes the power of the elements.

Darkness came o'er like chaos – and the sun
As startled with the terror seemed to run
With quickened dread behind the beetling cloud
The old wood sung like nature in her shroud
And each old rifted oak-tree's mossy arm
Seemed shrinking from the presence of the storm
And as it nearer came they shook beyond
Their former fears – as if to burst the bond

Of earth that bound them to that ancient place
Where danger seemed to threaten all their race
Who had withstood all tempests since their birth
Yet now seemed bowing to the very earth:
Like reeds they bent, like drunken men they reeled,
Till man from shelter ran and sought the open field.

Further reading

It is well worth reading poetry from an earlier time and if you are interested or involved in the natural world, John Clare's work may well appeal. Paul Farley has compiled a collection of Clare's poems, *John Clare: poems selected by Paul Farley* (Faber and Faber, 2007), which is a good introduction to his work.

The Letters of Pliny the Younger
by Pliny the Younger

Pompeii, near Naples, is the remains of a Roman town that was dev-
astated by the eruption of the volcano Mount Vesuvius in AD 79.
The town was covered in lava and ash, which helped to preserve
it, and lay buried for hundreds of years. Herculaneum was also dev-
astated and preserved, while Stabiae, mentioned in Pliny's letter, is
to be excavated and restored, and an archaeological park built.

Letters of Pliny the Younger to the Historian Tacitus:

*1. Pliny Letter 6.16 – Pliny the Younger describes the eruption and the
death of his uncle, Pliny the Elder, while trying to rescue survivors of the
early stages of the eruption.*

My dear Tacitus,

You ask me to write you something about the death of my
uncle so that the account you transmit to posterity[1] is as reli-
able as possible. I am grateful to you, for I see that his death will
be remembered forever if you treat it.[2] He perished in a devastation of the loveliest of lands, in a memorable disaster shared by
peoples and cities, but this will be a kind of eternal life for him.
Although he wrote a great number of enduring works himself,
the imperishable nature of your writings will add a great deal to
his survival. Happy are they, in my opinion, to whom it is given
either to do something worth writing about, or to write something worth reading; most happy, of course, those who do both.
With his own books and yours, my uncle will be counted among
the latter. It is therefore with great pleasure that I take up, or
rather take upon myself the task you have set me.

He was at Misenum in his capacity as commander of the
fleet on the 24th of August, when between 2 and 3 in the

---

[1]**posterity**   future generations
[2]**if you treat it**   Tacitus was to write about the eruption of Vesuvius in his
    Histories (although if he did, the book was lost)

afternoon my mother drew his attention to a cloud of unusual size and appearance. He had had a sunbath, then a cold bath, and was reclining after dinner with his books. He called for his shoes and climbed up to where he could get the best view of the phenomenon. The cloud was rising from a mountain – at such a distance we couldn't tell which, but afterwards learned that it was Vesuvius. I can best describe its shape by likening it to a pine tree. It rose into the sky on a very long 'trunk' from which spread some 'branches.' I imagine it had been raised by a sudden blast, which then weakened, leaving the cloud unsupported so that its own weight caused it to spread sideways. Some of the cloud was white, in other parts there were dark patches of dirt

and ash. The sight of it made the scientist in my uncle determined to see it from closer at hand.

He ordered a boat made ready. He offered me the opportunity
of going along, but I preferred to study - he himself happened to
have set me a writing exercise. As he was leaving the house he was
brought a letter from Tascius' wife Rectina, who was terrified by
the looming danger. Her villa lay at the foot of Vesuvius, and
there was no way out except by boat. She begged him to get her
away. He changed his plans. The expedition that started out as a
quest for knowledge now called for courage. He launched the
quadriremes[3] and embarked himself, a source of aid for more
people than just Rectina, for that delightful shore was a popu-
lous one. He hurried to a place from which others were fleeing,
and held his course directly into danger. Was he afraid? It seems
not, as he kept up a continuous observation of the various movements and shapes of that evil cloud, dictating what he saw.

Ash was falling onto the ships now, darker and denser the closer
they went. Now it was bits of pumice, and rocks that were black-
ened and burned and shattered by the fire. Now the sea is shoal;[4]
debris from the mountain blocks the shore. He paused for a
moment wondering whether to turn back as the helmsman urged
him. 'Fortune helps the brave,' he said, 'Head for Pomponianus.'

At Stabiae, on the other side of the bay formed by the gradually curving shore, Pomponianus had loaded up his ships even
before the danger arrived, though it was visible and indeed
extremely close, once it intensified. He planned to put out as
soon as the contrary wind let up. That very wind carried my
uncle right in, and he embraced the frightened man and gave
him comfort and courage. In order to lessen the other's fear by
showing his own unconcern he asked to be taken to the baths.
He bathed and dined, carefree or at least appearing so (which is
equally impressive). Meanwhile, broad sheets of flame were
lighting up many parts of Vesuvius; their light and brightness

---

[3]**quadriremes**   Roman galley ships with four banks of oars
[4]**shoal**   shallow

were the more vivid for the darkness of the night. To alleviate people's fears my uncle claimed that the flames came from the deserted homes of farmers who had left in a panic with the hearth fires still alight. Then he rested, and gave every indication of actually sleeping; people who passed by his door heard his snores, which were rather resonant since he was a heavy man. The ground outside his room rose so high with the mixture of ash and stones that if he had spent any more time there escape would have been impossible. He got up and came out, restoring himself to Pomponianus and the others who had been unable to sleep. They discussed what to do, whether to remain under cover or to try the open air. The buildings were being rocked by a series of strong tremors, and appeared to have come loose from their foundations and to be sliding this way and that. Outside, however, there was danger from the rocks that were coming down, light and fire-consumed as these bits of pumice were. Weighing the relative dangers they chose the outdoors; in my uncle's case it was a rational decision, others just chose the alternative that frightened them the least.

They tied pillows on top of their heads as protection against the shower of rock. It was daylight now elsewhere in the world, but there the darkness was darker and thicker than any night. But they had torches and other lights. They decided to go down to the shore, to see from close up if anything was possible by sea. But it remained as rough and uncooperative as before. Resting in the shade of a sail he drank once or twice from the cold water he had asked for. Then came an smell of sulfur, announcing the flames, and the flames themselves, sending others into flight but reviving him. Supported by two small slaves he stood up, and immediately collapsed. As I understand it, his breathing was obstructed by the dust-laden air, and his innards,[5] which were never strong and often blocked or upset, simply shut down. When daylight came again 2 days after he

<hr>

[5]**innards**   intestines

died, his body was found untouched, unharmed, in the clothing that he had had on. He looked more asleep than dead.

Meanwhile at Misenum, my mother and I – but this has nothing to do with history, and you only asked for information about his death. I'll stop here then. But I will say one more thing, namely, that I have written out everything that I did at the time and heard while memories were still fresh. You will use the important bits, for it is one thing to write a letter, another to write history, one thing to write to a friend, another to write for the public.

Farewell.

Further reading

If you would like to read historical fiction about the destruction of Pompeii, you could try *Pompeii* by Robert Harris (Arrow Books Ltd, 2004), in which a Roman Engineer Marcus Attilius Primus is the main character, trying to survive as Vesuvius erupts.

Caroline Lawrence's novel for children, *The Pirates of Pompeii* (Puffin Books, 2004), which is part of the *Roman Mysteries* series, may also interest you. It is an adventure story set after the eruption of Vesuvius.

Nature Cure
by Richard Mabey

> Richard Mabey's biography *Nature Cure* (Vintage, 2008) describes his experience of depression and how his recovery was helped by his love of nature. It explores how we can live a simpler life, and is also a homage to the 19th century nature poet, John Clare.

Flitting

I dwell on trifles like a child
I feel as ill becomes a man
And still my thoughts like weedlings wild
Grow up to blossom where they can.
                    John Clare, 'The Flitting'

It's October, an Indian summer. I'm standing on the threshold like some callow[1] teenager, about to move house for the first time in my life. I've spent more than half a century in this place, in this undistinguished, comfortable town house on the edge of the Chiltern Hills, and had come to think we'd reached a pretty good accommodation. To have all mod cons on the doorstep of the quirkiest patch of countryside in south-east England had always seemed just the job for a rather solitary writing life. I'd use the house as a ground-base, and do my living in the woods, or in my head. I liked to persuade myself that the Chiltern land-scape, with its folds and free-lines and constant sense of sur-prise, was what had shaped my prose, and maybe me too. But now I'm upping sticks and fleeing to the flatlands of East Anglia.

My past, or lack of it, had caught up with me. I'd been bogged down in the same place for too long, trapped by habits and memories. I was clotted with rootedness. And in the end I'd fallen ill and run out of words. My Irish grandfather, a

---

[1]**callow**   lacking experience

day-worker who rarely stayed in one house long enough to pay the rent, knew what to do at times like this. In that word that catches all the shades of escape, from the young bird's flutter from the nest to the dodging of someone in trouble, he'd flit.

Yet hovering on the brink of this belated initiation, all I can do is think back again, to another wrenching journey. It had been a few summers before, when I was just beginning to slide into a state of melancholy and senselessness that were incomprehensible to me. I was due to go for a holiday in the Cevennes with some old friends, a few weeks in the limestone *causses*[2] that had become something of a tradition, but could barely summon up enough spirit to leave home. Somehow I made it, and the Cevennes were, for that brief respite, as healing as ever, a time of sun and hedonism and companionship.

But towards the end of my stay something happened which lodged in my mind like a primal memory: a glimpse of another species' rite of passage. I'd travelled south to the Herault[3] for a couple of days, and stayed overnight with my friends in a crooked stone house in Octon.[4] In the morning we came across a fledgling swift beached in the attic. It had fallen out of the nest and lay with its crescent wings stretched out stiffly, unable to take off. Close to, its juvenile plumage wasn't the enigmatic black of those careering midsummer silhouettes, but a marbled mix of charcoal-grey and brown and powder-white. And we could see the price it paid for being so exquisitely adapted to a life that would be spent almost entirely in the air. Its prehensile claws, four facing to the front, were mounted on little more than feathered stumps, half-way down its body. We picked it up, carried it to the window and hurled it out. It was just six

---

[2]**causses**   limestone plateaus in the Massif Central, a region in south-central France

[3]**Herault**   a region in the south-west of France

[4]**Octon**   a district in the Herault

weeks old, and having its maiden flight and first experience of another species all in the same moment.

But whatever its emotions, they were overtaken by instinct and natural bravura.[5] It went into a downward slide, winnowing[6] furiously, skimmed so close to the road that we all gasped, and then flew up strongly towards the south-east. It would not touch down again until it came back to breed in two summers' time. How many miles is that? How many wing-beats? How much time off?

I tried to imagine the journey that lay ahead of it, the immense odyssey[7] along a path never flown before, across chronic war-zones and banks of Mediterranean gunmen, through precipitous changes of weather and landscape. Its parents and siblings had almost certainly left already. It would be flying the 6,000 miles entirely on its own, on a course mapped out – or at least sketched out – deep in its central nervous system. Every one of its senses would be helping to guide it, checking its progress against genetic memories, generating who knows what astonishing experiences of consciousness. Maybe, like many seabirds, it would be picking up subtle changes in airborne particles as it passed over seas and aromatic shrubland and the dusty thermals above African townships. It might be riding a magnetic trail detected by iron-rich cells in its forebrain. It would almost certainly be using, as navigation aids, landmarks whose shapes fitted templates in its genetic memories, and the sun too, and, on clear nights, the big constellations – which, half-way through its journey, would be replaced by a quite different set in the night sky of the southern hemisphere. Then, after three or four weeks, it would arrive in South Africa and earn its reward of nine months of unadulterated, aimless flying and playing. Come the following May, it and all the other first-year birds would come back to Europe and race recklessly

---

[5]**bravura**   skill and artistry
[6]**winnowing**   being tossed in the air
[7]**odyssey**   series of travels

about the sky just for the hell of it. That is what swifts do. It is their ancestral, unvarying destiny for the non-breeding months. But you would need to have a very sophisticated view of pleasure to believe they weren't also 'enjoying' themselves.

When that May came round I was blind to the swifts for the first time in my life. While they were *en fête* I was lying on my bed with my face away from the window, not really caring if I saw them again or not. In a strange and ironic turn-about, I had become the incomprehensible creature adrift in some insubstantial medium, out of kilter with the rest of creation. It didn't occur to me at the time, but maybe that is the way our whole species is moving.

Further reading

You could take a look at Richard Mabey's book for children, *Oak and Company* (Pan, 1999), which is beautifully illustrated and describes the life of the tree and the animals and plants that frequent it.

Activities

Dust Storm

Before you read

1   Imagine a dust storm that towers above buildings, causes breathing difficulties and is so dense that it is impossible to see. Think of several powerful verbs to describe it and write them down.

What's it about?

Read the text, then work with a partner to make short notes in answer to questions 2 to 4.

2   Can you guess in what country the story is set? How? Study the language as well as the setting.

3   Why does the narrator get caught up in the storm? Where is she? What journey does she have to take?

4   Which are the most startling examples of the dust storm's destructive power? In what ways is the narrator's life at risk?

Thinking about the text

5   This narrative poem has a simple direct style. However, we have a strong impression of the storm and its effects. From time to time, powerful verbs such as 'bowed down' (page 100) and images such as 'voices rose like ghosts on the howling wind' (page 101) are used. Find more examples of powerful verbs and images.

6   a   The narrator waits up all night for her father to return, but there is no sense of panic. There is fear, however. How is it conveyed? Find a striking example.
    b   Write the narrator's diary entry for that night, in a non-dramatic style, but revealing her fear and worry.

7   Work in a small group to improvise a short drama about the storm and the events in the poem. Include:
    ● the characters of Billie Jo and her father
    ● characters who bring news, for example Joe De La Flor.

    Decide:
    ● where your improvisation will be set (e.g. indoor or outdoors)
    ● how you will convey the characters' experiences of the storm
    ● how you will use body language as well as dialogue.

Hurricane Gold

Before you read

1 What films have you seen in which a disaster occurs? Were you convinced by the special effects? Did you identify with the characters in their predicament? Jot down your thoughts.

What's it about?

Read the text, then work with a partner to make short notes in answer to questions 2 to 4.

2 Who are the intruders looking for? Who do they capture and what do you think the purpose of this might be? How does James react?

3 How does James escape? Trace his movements from the beginning to the end of the text.

4 Do you think the hurricane helps James or hinders him, or both?

Thinking about the text

5 The writer creates a strong sense of the hurricane's presence by:
   - referring in detail to the damage it causes (such as the force of the wind on the house on page 108)
   - using descriptive language (such as the series of adjectives on page 105: 'crashing, hissing, roaring, squealing, rumbling').

   Find more examples of these techniques and choose the one you like best. Use it as a model for a paragraph of your own, in which you describe the hurricane from James's point of view.

6 What do you think has happened to JJ and Precious? Draw a spider diagram with their names in the centre. Think of as many possibilities as you can and record them on your diagram. You could use one of your ideas to write a story.

7 Imagine that James has to send a report for M to include details of:
   - the nature of the hurricane and the extent of the damage
   - what happened to Precious and JJ during the hurricane
   - the course of action taken by James and why.

   Write his report, using a formal style. Include place, and time. You could begin like this:
   Lagrimas Negras, 14.00 hours, Agent Bond reporting. Incident at Tres Hermanas.

To Build a Fire

Before you read

1 Imagine it is a cold, wintry day with a piercing wind and a threat of snow. You are cold and tired and making your way home. Write a few sentences to describe your feelings.

What's it about?

Read the text, then work with a partner to make short notes in answer to questions 2 to 4.

2 Do you know where the Yukon is? Make a note of information from pages 109–110 that tells you and, if you can, find a map of the region.

3 'The landscape and climate are as important as the main character in the story.' Do you agree with this statement? Why or why not?

4 In what sense has the main character pitted himself against nature? Who wins?

Thinking about the text

5 The story opens: 'Day had broken cold and gray, exceedingly cold and gray . . . ' What kind of mood is immediately created? Find further examples in the first few paragraphs that:
   - reinforce this mood
   - suggest a warning
   - convey the nature of the landscape and the climatic conditions.

6 Make a list of information about the main character, thinking about:
   - why he had undertaken the journey
   - his experience
   - his strengths and weaknesses.

7 a With a partner, discuss the following statement from the story and its importance:

   Its [the dog's] instinct told it a truer tale than was told to the man by the man's judgement.

   b Why is the husky dog an important feature in the story? What does it tell us about the man? Discuss your ideas in your group.
   c Present your ideas to another pair. Then try to reach agreement about the importance of the dog in the story.

8 If this story gives advice about life, what is it?

Wind

Before you read

1   Find out about the Great Storm of October 1987. Why did it occur? What damage and loss of life was caused? With a partner, do some research on the Internet.

What's it about?

Read the text, then work with a partner to make short notes in answer to questions 2 to 4.

2   Where do you think the house in the poem is situated? What kind of house is it? Record words or lines that help the reader to build a picture of the house and its surrounding features.

3   What is the time scale of the poem? What time of day or night does it pass through? What vivid impression are we left with at the end?

4   Discuss what is happening in the second verse. Read each line carefully, deciding what impression we form of the hills and what the colours represent.

Thinking about the text

5   In the poem, the wind is so strong that even the birds are affected:

The wind flung a magpie away and a black-
Back gull bent like an iron bar slowly

The alliteration (using the letter 'b') helps to create a wind's robust, vigorous sound. Create images to suit the following in a storm:

ivy clinging to a wall   refuse bins   a wooden gate

6   **a**   Re-read verse three. The narrator of the poem is depicted in an original and startling way. How does he describe the strength of the wind? What does it seem might happen to the hills at any moment?

   b   Write a prose poem about yourself in a high wind or storm. Write in the usual way as prose, not in lines as a poem, but include images and description. Do not write a story. To start you off, either:
   - focus on a verse from 'Wind' and conjure up a detailed picture in your mind, *or*
   - quote a line from the verse as a starting point.

The Hail Storm in June 1831

Before you read

1 Think back to your earliest memories. Were you afraid of storms as a young child? Did you hide from thunder and lightning or did you watch it from the window? Share your memories with a partner.

What's it about?

2 How does the storm begin? What do you associate with the words 'chaos', 'terror', 'dread', 'shroud?' Why has the poet chosen such words to describe nature? Jot down some ideas.

3 Read the last line. Why would the people give up their shelter and run into the open? Discuss your ideas with a partner, choosing the most likely explanation.

4 Read line seven. Today we would say 'And as it came nearer . . . ' Find other examples of words or expressions that tell you that the poem is set in the past.

Thinking about the text

5 The poet personifies the sun (makes it seem like a person), describing it, 'As startled with the terror', as though the sun had an expressive face. This helps us to identify with the experience of the storm.
   a Record other examples of personification from the poem.
   b Create some images of your own in which the storm is personified.

6 Draw an illustration of the scene, without using colour. Include the effects of the storm on:
   - the wood, thinking about the shape of the trees
   - the sun
   - the clouds.

   Write a poem to go with your illustration. You could use personification if you wish, using the work you did in question 5.

7 Imagine you are the poet. You always carry a notebook with you to jot down ideas for poems. Jot down the notes you made when you witnessed the storm.

The Letters of Pliny the Younger

Before you read

1 What do you know of Pompeii and Herculaneum? Why are they remarkable? Share your ideas and experiences in a group.

What's it about?

Read the text, then work with a partner to make short notes in answer to questions 2 to 4.

2 Who is Pliny the Younger writing to and why? Think of more than one reason.

3 Give details of Pliny the Elder's status and interests. Where was he at the time of the eruption and when did it occur?

4 Find a map of the region or visit http://en.wikipedia.org/wiki/Pompeii, which gives details of the area. Note where Vesuvius, Misenum and Stabiae are, and the route taken by Pliny the Elder.

Thinking about the text

5 Study the letter for information. Draw a series of diagrams or sketches showing the pattern of the eruption from the first details to the last. Label your diagrams as necessary and write a caption under each one explaining what is happening.

6 Work with a partner to carry out a hot-seating exercise, in which one of you takes the role of Pliny the Elder and the other is an interviewer. Focus your questions and answers on:
- Pliny's initial reasons for setting out towards Vesuvius
- why he altered his plans
- his awareness of danger.

7 Write a eulogy for Pliny the Elder (a eulogy is a speech in praise of someone, usually when they have died). Note the vocabulary and expressions used by Pliny, such as:

for I see that   perished   enduring
Happy are they   It is . . . with great pleasure

Try to adopt the same style and tone – formal and respectful but also written with warmth.

Nature Cure

Before you read

1 The aurora borealis or starlings flocking are often quoted as examples of spectacular natural events. What are they? Check in an encyclopaedia. Jot down your answers.

What's it about?

Read the text, then work with a partner to make short notes in answer to questions 2 to 4.

2 Why did the Chiltern landscape once seem so important to Mabey?

3 What do you think has happened to the writer? How does he feel?

4 Re-read the last line of the text (page 139). In what sense is it a warning to us all?

Thinking about the text

5 Re-read the verse from John Clare's *The Flitting* (page 136). What emotions and thoughts are expressed about moving home? Why has Richard Mabey chosen it to open his book? Discuss your ideas with a partner and make notes. Then find the poem and read it together.

6 a Richard Mabey's writing is often poetic. He creates images, mainly of nature, which give impact to the feelings expressed. What do the following convey about his situation?
   - bogged down (page 136)
   - clotted with rootedness (page 136)
   - the young bird's flutter from the nest (page 137)

   b Write a descriptive paragraph (or a poem) about being trapped by circumstance and having dreams of escape.

7 One of the most emotive parts of the text is the fledgling swift's struggle to live and the extraordinary journey it is about to take. Work with a partner to present a reading of a section of the text, from 'But towards the end of my stay' (page 137) to the end.
   - Study the images and sounds, and practise reading them with expression. (For example, 'beached in the attic' emphasises that the bird is out of its element, the air. You could say the words more slowly to accentuate the image.)
   - Decide where else to change pace and voice (e.g. questions encourage a natural change of voice).

Compare and contrast

1   The texts give varying impressions of nature's power and its effects. Which do you find:
   - the most startling
   - the most frightening
   - the most moving?

   Discuss your choices with a partner.

2   Both John Clare and Ted Hughes wrote poems about the wind. Write an essay comparing and contrasting the two poems. First draw up a 'Similarities and Differences' table and make notes, thinking about:
   - the images created
   - the setting of the poems
   - the techniques used (such as personification).

3   Which text do you like best and why? What qualities does it have? Choose the one you like best in this section and carry out an interview for a website, in which you ask the writer how they came to create their work. Assemble a selection of open and closed questions. Open questions are those with no definite answer, and which allow the speaker to express ideas and feelings. Closed questions usually have one answer. (An example might be 'Do you write every day?')

4   The texts in this section explore the power of some of nature's contrasting elements: fire, ice and wind. Design a wall-frieze for an important building, such as a museum, that show these contrasts. Think about:
   - the power of each element
   - the choice of colour and patterns.

   Then provide a commentary on your frieze saying how it conveys the power of each element. Include a quotation about each element from three suitable texts.

5   This section includes examples of work by Richard Mabey and John Clare. How are their themes similar? In what ways do you think the latter has influenced the former? Discuss these points in a group, referring to their work to support your ideas.

4 Future planet

The previous sections of this anthology have looked at the remarkable variation and beauty of our world, largely seen through the eyes of naturalists or those who have been touched by the power of nature. This final section of *The living planet* turns to the greatest challenge that is likely to face us all in the coming decades – Earth's changing climate and how to reduce its effects and manage resources. It includes:

- imaginative accounts of a future world
- a celebration of a species we have lost
- information about how we use and misuse the Earth's resources
- questions and answers about what climate change means
- information about what we can do to take care of the planet.

The last point is the one that should be at the top of our list. You may like to think about the different ways you can take care of the planet and what you are already doing.

Activities

1   Fair trade means paying a fair price to people in developing countries for goods such as coffee, tea, cocoa, bananas, fresh fruit and crafts. Find out about fair trade at www.oxfam.org.uk/coolplanet/kidsweb/food.htm. Then create a colourful leaflet outlining the main points about fair trade. Include a section that shows how these goods are produced without harming the natural world.

2   In 2008 there was around 6.7 billion people on Earth. By 2042 there are likely to be nearly 9 billion. We all need food, and land to grow it. Lack of food has been a fact of life for people across the world for many years. Global food shortages are increasingly likely. Find out more about this: go to http://www.oxfam.org.uk/education/resources and click on Global Food Crisis. Then work in a small group to present the data you have found in a school assembly. Discuss how you will do this effectively. For example, you could use a flip board to present facts and figures.

3 **a** In the introduction to *The Rough Guide To Climate Change*, James Lovelock says that 'the era of cheap, plentiful oil is drawing to an end' and discusses how this could lead to a better Earth. In a small group, discuss:
   - what we use oil for
   - how the lack of oil might lead to a cleaner planet
   - what different kinds of sustainable energy there are.

 b Visit http://www.ashdenawards.org/schools. This site gives information on schools that use sustainable energy. Look at the technologies and the case study database. Then, as a group, choose a suitable form of sustainable energy and make a case for its use by your local community.

Green Boy

by Susan Cooper

Green Boy is set in the Bahamas, where the wildlife is under threat from development. Two local brothers, Trey and Lou, enter the parallel world of Pangaia, where the natural world is also being destroyed and dangerous mutations have evolved. While their grandfather fights the changes to the island, Trey and Lou try to save Pangaia.

The forest was thick and lush, with high bushes and leafy vines tangling between the trees, and giant ferns and mosses mounded over everything on the ground. Rocks, earth, fallen branches or trees – they were all turned bright green, all swallowed by growing things. Yet it wasn't like any forest I'd ever seen; there was something very spooky about it. Not a glimmer of sunlight came down through the high canopy of branches and leaves; the light was dim, and the air was thick and still and humid. After that one horrific yellow moth, there were no insects to be seen, no butterflies or beetles or wasps or flies, and no birds either. You could hear harsh cries and croakings out there in the trees, but nothing flew or fluttered through the air, or moved on any branch.

As I looked more closely, I began to see that all the trees were exactly the same kind of tree, great spreading giants with broad roots and strange scaly bark, and that the vines were all one kind of vine, thousands of clambering, twining stems, thick as a man's leg and sprouting clusters of broad round leaves. The ferns were all alike too: tall arching fronds as high as my head, with yellow-green leaf-divisions like fingers, the back of each one of them studded with those brown spore cases you see on most ferns. I put out a finger to touch one, and the thing vanished in a puff of floating brown dust. It smelt bitter, like vinegar.

And Lou walked through all this as if he were following a route he had known all his life.

'The tree will call him,' Bryn had said, and what did that mean, for goodness' sake? The place was full of trees, and trees don't talk. But from the moment those words came out of Bryn's mouth, my little brother Lou stood still, and stiffened, and seemed to forget that anyone else was there. Even me.

He started to move ahead through the forest, slipping through ferns and around trunks and low branches, with such silent certainty that we all followed him. I stayed close behind him, and soon I realised that he had taken us to a kind of path, a clear way through the thick tangled growth that showed no sign of having been cut or cleared, yet was wide enough to let a full-grown man pass through.

Math was at my heels; I glanced over my shoulder and saw him looking ahead at Lou intently, out of his bright dark eyes. I also saw, for the first time, that the handle of a big knife was sticking out of the top of one of his close-fitting boots. The boots were still soaked with water; it was the faint squishing sound that had made me look down at them.

In that same moment, everything around us erupted. Lou leapt back, yelping, Math's hand flashed down and up, and the knife was flying through the air to bury itself, ahead, in something black and furry and snarling, blocking the path.

Lou fell, and I grabbed him. Bryn and Annie were diving forward, past me. 'Get back!' Annie hissed at me. 'Keep him back!' And I saw they had knives too, and were striking like Math at the thing ahead, again and again, grunting, intent.

The thing screamed, a horrible high shriek, and then it was silent and still.

Math brought his knife down in one last violent slash, and a spray of liquid came flicking back at us. When I wiped it off my arm I saw that it was blood.

The creature looked like a rat, the size of a dog. It had a pointed nose and a long hairless tail, and its body was sleek black. It lay there, filling the pathway. The jaws were stretched wide and menacing, full of huge sharp yellow teeth. There was blood everywhere. I held Lou close; I knew he must have been terrified.

But Lou was wriggling out of my hands, not shaking, not trembling on the edge of a seizure, not even visibly afraid. He was looking ahead. He had his head up as if he were listening. He gave a little gurgle that was almost like his happy sound, and he kept on going, walking round the path and the twitching body of the giant rat.

'Lou!' I shouted.

'Just follow him,' Annie said in my ear. 'He can hear the tree.'

'*What tree?*' I said unhappily.

The others were passing us, following Lou. Annie held me back. 'Listen to me,' she said. 'In your world, Lou is nothing, but in ours, he is magical, he is predestined. We have been waiting for him. Only he can save this world, only now and only here.'

I couldn't make sense of any of this. 'I have to look after him! I always have – he's my little brother, it's my job.'

'It's still your job,' Annie said. 'But it's ours too.' She gave me a big warm smile, that lit up her face like sunlight, and tugged my hand so that we were running together to catch up with the rest. The shrieking in the trees grew suddenly louder, and in a great flapping flurry a big dark-coloured bird swooped low over our heads and away again, into a tree. I couldn't see it clearly. It looked a bit like a golling, a bird we have in the islands that's the size of a duck, and usually comes out only at night. But this one was far, far bigger.

'It's harmless,' Annie said. 'Keep going.'

'They're all huge. Everything.'

'Mutants.' We'd caught up with the others; I could see Lou trotting purposefully ahead of them. 'That's why Government preserves the Wilderness – to study them. There's a big research facility somewhere in here, and the whole area's shut off, and guarded. It's a bad place, the Wilderness. If it didn't have some of the oldest trees on the planet, we'd have attacked it long ago.'

With a swoosh, the dark bird flapped back across the path over our heads, and then into the trees again. It didn't seem able to fly very high. Bryn and Math looked up uneasily, but little

Lou paid no attention; he went on, without pausing. The forest seemed a bit brighter here, as if more light were filtering down through the thick criss-cross of branches overhead. There were fewer trees too. Gradually, as we went on, the light grew, and I began to see a few chinks in the green ceiling, a few glimpses of hazy sky. We were beginning to come out of the forest.

And then, in an instant, we were out of it, and all of us stopped, bumping into each other like a line of dominoes.

We were standing on a slope, there among the few remaining trees, and spread before us was a grey world of concrete and steel and stone, an unbroken city, stretching as far as we could see in all directions. Buildings and streets filled an enormous plain, flat, flat, until in the distance it rose into slopes like the one on which we stood. Beyond them, folds of hills rose higher and higher, each of them grey with the buildings of the city, until they vanished into a brown haze.

Gwen was standing beside me now, looking out. 'There is what we've done to Pangaia,' she said bleakly. 'There's our world.'

Lou had paused only for a second. He was standing there now on the mossy slope, turning his head from side to side, like a dog casting about for a scent. Then he began to move again, sideways, towards a rocky outcropping surrounded with the same tall ferns that had crammed the floor of the forest. And I saw that beyond the rocks, between the forest and the endless sprawl of the city, a tremendous steel-mesh fence stood as a barrier, topped with whorls of barbed wire. Beyond it was a second fence, just as big, just as sturdy, and beyond that a third. If Lou's strange convinced sense of direction was leading us all towards the city, we'd have a hard time getting there.

We followed Lou, walking now over stony ground patched with moss and clumps of an odd yellow grass. I was staring out at the fences, which looked taller and more forbidding the closer we got to them. I said to Gwen, 'How can we possibly get out of here, with those in the way?'

She said, 'If the tree is inside the fence, just be grateful we're here. It would have been even harder to get in than to get out.'

There was that word again. *What tree?* I said.

Then gradually I began to hear, somewhere, a sound that seemed to come straight from Long Pond Cay: a weird husky whistling sound like the wind in the casuarina[1] trees. It was soft but unmistakeable, and though it didn't grow louder, it didn't go away. It seemed to fill the air all around us; I couldn't tell where it was coming from.

Lou walked round the group of tall irregular rocks ahead of us on the slope, and as we followed him I could see an inner cluster of rocks with a tree growing out of them. Its roots were spread over one flat rock like long dark fingers. It wasn't very

---

[1]**casuarina**   a tree or shrub with jointed stems and leaves like scales

big, maybe twenty feet high, but it looked very, very old; its trunk was broad and twisted, with grey bark smooth as stone, and its thick, gnarled branches drooped, as if they had carried a heavy weight for a long time. They were covered all over with new side shoots and twigs bearing long thin leaves almost like pine needles. And all these were moving in a breeze that I couldn't feel; a breeze that came from nowhere, and blew only on this one tree, producing that soft moaning casuarina sound that was filling the air. It was like singing, though it had no words.

Perhaps it had words for Lou. He paused, and clambered up on to a ledge of rock on the way to the tree. Bryn and Math climbed up after him – and then they both gave a kind of strangled gasp, and grabbed fast for their knives, staring downward. Gwen and Annie and I scrambled up to look over the edge of the rock.

It was like a snake pit, all round the tree. Between the outer group of great lichen-patched boulders where we were standing, and the rocks in the middle where the tree grew, there was a gap like a big ditch, filled with shiny black bodies, moving, slowly rippling, like a sluggish sea. They were hideous: thick armour-plated cylinders about three feet long, with tiny snout-like heads. At first they looked like stubby hard-shelled snakes, but after a moment you could see the legs, hundreds of legs, moving ceaselessly under each body. There was a very faint smell, that reminded me of something, though I couldn't remember what.

Bryn was looking down in fascinated disgust. He took a breath, and I saw his fingers tighten round his knife. He said to Math, 'If we go down together, back to back, we can cut a way through.'

But Math didn't have a chance to answer. Lou swung round in front of them, making the throaty noises that for him were 'No, no!' He was shaking his head violently, and he had his hands up, trying to push them both backwards, away from the ditch.

Math looked down at him in astonishment. He said, 'What, then?'

Lou put his finger to his lips. He held up the other hand with its palm towards them. Then very slowly he began to edge himself down into the pit full of those squirming black monsters.

'Ah, no!' Math said, appalled. He started forward, and Lou stopped, frowning, and held up his hand again.

'Let him be,' Bryn said. 'He knows what he wants to do.'

And just as I began to panic, I remembered in an instant where I had first smelled the faint smell that was in the air here. It was the tiny hint of a scent that you could catch once in a while from one of Lou's favourite playthings at home on the island: the little black millipedes, that grossed out Grammie but curled obligingly into a harmless circle in Lou's gentle fingers.

I stared down at the pit. These awful-looking things, long as my leg, were gigantic mutant versions of Lou's millipedes. He seemed to have recognised them at once – but would they recognise him? I had a sudden terrible vision of him down there, screaming, covered with flailing black bodies.

Lou looked up and caught my eye, and shook his head. He grinned. He knew just what I'd been thinking.

Then he climbed slowly down into the ditch, and squatted at the edge of the mass of black bodies. He reached down and patted one of them; then knocked on its hard back with his knuckles, and laughed.

The millipede curled itself slowly into a circle. So did the next, and the next. And on, and on, until every one of them was curled up like an automobile tyre, lying there unmoving, unthreatening, in a harmless heap.

'I'll be damned,' Bryn said.

Lou laughed again, and he walked across the ditch, stepping lightly from one black curled body to the next, until he reached the other side.

Further reading

You could find out what happens to Trey and Lou and their island by reading the complete novel *Green Boy* (Aladdin Paperbacks, 2003).

If you like Susan Cooper's style you could try reading her historical novel *King of Shadows* (Aladdin Paperbacks, 2005), which is set in the Elizabethan period and features William Shakespeare.

Floodland

by Marcus Sedgwick

Floodland is set in a future in which much of Britain's coastal areas have flooded. It tells the story of Zoe, who sets out from Norwich in a small boat. This is the beginning of the story.

Zoe ran. Harder than she had ever run in her life. Her feet pounded through the deserted streets of derelict buildings. Somewhere, not far behind, she could hear the gang coming after her. It felt as if her heart would burst, but she didn't slow down. She'd been planning to leave the island for a long time, but had been putting it off. It was a big decision to set out to sea in a tiny rowing boat. Now she had no choice.

Before, no one had bothered her. Zoe was a loner. Most of the people left on Norwich hung around together in groups, but she preferred to be on her own. It was safer that way, because you never knew whom you could trust.

Somehow, someone had found out about the boat she'd been hiding. A boat was an escape route, a way to get away from Norwich, which got smaller every year, as the sea kept on rising. It didn't matter that there could only be room for two people at most in her boat. Others had joined in the chase, and now a mob of about fifteen people was hot on her heels. There was only one way out; to get to her boat before they got to her. So she ran on, while her body screamed for her to stop.

'Get back here!' someone yelled angrily at her, though they couldn't see her.

It wasn't far to the little shed where she'd hidden *Lyca*, her boat. A couple more streets of derelict shops to where what was left of the city fell away into the sea. If the sea hadn't come she might have been shopping here herself, with her parents perhaps. From much practice she squashed the thought of her parents as soon as it started, and kept on running.

Just before she rounded a corner, she heard more shouts from behind. They had seen her.

'There!'

'Come on!' shouted another voice. 'Get her!'

Scared, she made the corner, but her feet slipped from under her on the wet ground. She went sprawling, and slid clumsily in the mud. She started to panic badly, and made a mess of getting up again. She had dropped her pack as she fell, but there was no time to pick it up.

The sound of running feet came closer. Another two seconds and they would be round the corner. She got up and practically threw herself over a wall. She landed awkwardly, but she'd won a little more time. She was in a graveyard. It led away down a hill to where a small brick shed stood near the water's edge. Once it had contained all the equipment for looking after the graveyard, but now it contained Zoe's boat. The previous night she had rowed around from the warehouse where she had found the boat and fixed it. The old building had been unsafe when she'd discovered it, and had been getting worse. She had decided to find a new place to keep her boat, and the shed seemed ideal.

In the dark she had dragged the boat the short distance from the water to the shed. It had been very hard work. At night she hadn't noticed the deep ruts the boat's keel had made in the sodden grass. In daylight, even in her mad rush, they were obvious. She would be lucky if no one had already found it.

'*Lyca*,' Zoe panted as she opened the shed door, 'please be here, *Lyca*.'

It was all right. The boat was still there waiting for her.

Pulling it across the grass, and then into the water, she dared to look behind her for the first time. Her stomach twisted with fear. The gang were storming down the hill, weaving in and out of the crumbling gravestones. Zoe moved faster. She clambered aboard and put the oars out, then started to pull. They were at the water now, and though one or two stopped, the rest came splashing madly after her.

'Take me with you!'

'Come back! We won't hurt you. Just take us with you!'

Zoe could see their eyes, clearly. She saw fear. But she couldn't trust them. Since she'd lost her parents, she'd made it a rule not to trust anyone. Zoe had heard people say they'd lost someone, when really they meant they had died. In Zoe's case, 'lost' meant exactly that. It was still unbelievable, and so stupid.

She looked at the crowd in the water again. If she went back, there'd be a fight over her boat, and she wouldn't get a look in. She rowed on, pulling harder, even though she was safely away.

Slowly, she watched as the people dragged themselves out of the water and waded back to the shore. Natasha was there too. That hurt most of all. Natasha was the closest thing she had to a friend. Zoe used to see her when the supply ships came, before they stopped coming. After that she saw her sometimes at the allotments, when she went to put some work in to earn food. They would only have a little chat, but it was enough to keep Zoe from cracking up. But now the allotments had sunk into chaos, too.

Zoe suddenly remembered their conversation the last time they'd met. She had been about to tell Natasha about her boat, and her plans to escape, but had decided not to. Maybe Natasha had guessed? From something Zoe had let slip? It didn't matter now. The crowd stood quietly, watching her as she rowed away.

Zoe didn't feel scared of them any more.

'Sorry,' she said to herself, quietly. She began to cry, but she didn't stop rowing. Her uncut hair fell across her eyes, but she didn't stop to push it away. Still she rowed on, her thin hunched frame working the oars until finally she had to pause for breath.

Feeling around in her pocket she fished out her compass. It was the last thing she owned that had belonged to her parents. For that reason she'd kept it in a pocket. If she hadn't she'd have lost it when she dropped her pack. It was a little dented from her leap over the cemetery wall, but it was still working.

She pointed herself south-west, and rowed. She couldn't remember the name of the place the supply ship used to come from, but she knew the big bit of Britain was somewhere in that direction.

She was rowing away from all she had ever known. It was a strange thing. Before the previous night, she had only ever pretended to row. Her dad had taught her, in the same methodical way he did everything.

'You'll need to know how to do this one day,' he told her.

He'd taught her how to use the compass, as well as a lot of stuff about survival. Just in case the time came when she was on her own.

And so every now and then, when they weren't busy just trying to get by, they'd sit in an old bathrub and pretend to row.

Even though it had seemed like a game to Zoe at the time, he'd made sure she was doing it right anyway. And she knew just how to do it, the only thing that surprised her was how hard it was to pull the oars through the water.

'Why don't you look where you're going?' she'd asked her dad.

'When you're rowing, you mean?' he said.

'Yes. Why do you sit looking backwards?'

'It's just the way it's done,' he said. 'You couldn't row half as well facing forwards.'

It had always seemed strange to her, but now it was even worse. There before her was Norwich getting smaller and smaller with each stroke. She was heading into the unknown, without even looking where she was going.

She rowed and rowed, until her small supply of food had gone. She had put the compass on the floor of the boat in front of her, and every few seconds she checked her direction against it. There was no sign of land now, and a creeping fear began to seep into her. She looked at the compass almost every stroke; it was her only chance now. Like magic, its tiny hand kept pointing in the same direction. It knew where she was going, even if *she* didn't. She lost all sense of time. The sun was somewhere way overhead, and beat on the back of her neck, making her feel dizzy. She pushed her hair out of her eyes, but the sea wind blew it back across her face. She felt faint. She was in trouble. She had just enough awareness to pull in her oars. Then she slumped over them.

In her stupor she replayed the nightmare where she had lost her parents.

Further reading

If you enjoyed the first chapter from *Floodland* (Dell Yearling, 2002) you may like to complete the novel. Marcus Sedgwick also drew the illustrations for the book.

Sedgwick has written several other novels that you might like to read. His two latest are *My Swordhand is Singing* (Orion Children's, 2006) and its sequel *The Kiss of Death* (Orion Children's, 2008).

Two poems

by Jamie McKendrick

The poem *From the Flood Plain* and its companion piece *Après* look at the effects of flooding from a witty and unusual point of view.

From the flood plain

No flood as parched as this – a mere foot
or two of gilded bilge – will turf us out
from the lands of the frog and the newt
who for the best part of a century
have bided our time in the tall grass.
We've stood their ground, and stand it still,
though our legs are cased in long green boots
and the sofa's propped on a tower of bricks.

Unmoved, we see fish swim in the back yard
and a swan sneer from the vicar's garden,
though the cold waters still keep rising,
working away at the silicone seals,
unpicking the doors we've turned into dykes
and days may pass before our power's restored.

Après

When the flood waters left they left
the pine boards cupped; the plaster blistered
with salts; the cheap chipboard
bursting out of its laminate jacket
in all the kitchen units; the electrics wrecked
with the wires firing in the sockets;
the polyfilled cracks in the buckled doors
once more agape; the iron grate sporting
a hem of rust and the ash it contained
arranged in a scum-line above the skirting;
dampness, months-deep, fattening the pores

of the brickwork; a question-mark over the slate floor;
the oven fouled; the fridge unsafe; the whole place
humming with marsh rot and fetor[1]
but the garden, the garden good, and greener
for an alien crop of hogweed higher
than us, hardy, sturdy, hirsute,[2] armed
with a poison sap against expulsion.

Further reading

Seamus Heaney's poem *Bogland*, which you can find in *Door into the Dark* (Faber and Faber, 2002), also gives us a striking description of how nature can flourish, this time in a Irish setting. Or, for a dramatic poem about wild waters, read John Clare's *The Flood*, in *John Clare: poems selected by Paul Farley* (Faber and Faber, 2007).

---

[1]**fetor**   strong, unpleasant smell
[2]**hirsute**   hairy, bushy

Classic Combo

by David Heatley

Graphic fiction is similar to a comic. It tells a story or presents ideas through illustrations and cartoons. It may be lengthy, as in the case of the graphic novel, or short like *Classic Combo*.

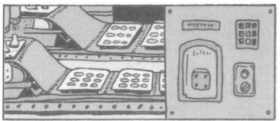

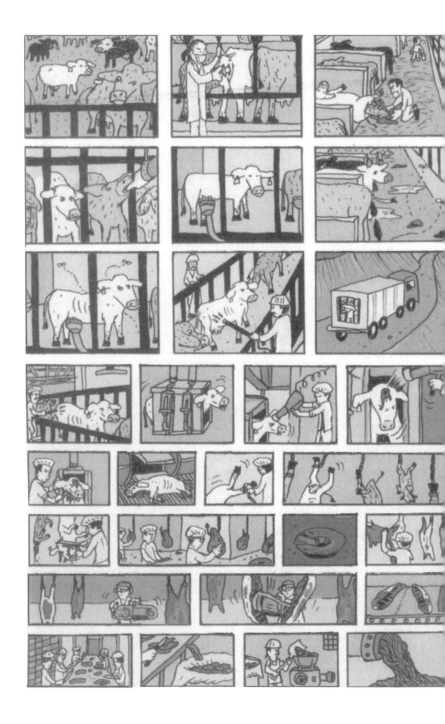

COLORING BY DAVE KIERSH   DAVID HEATLEY 2008

Further reading

If you enjoy graphic fiction, you may also enjoy animation. Hayayo Miyazaki's award-winning film *Spirited Away* (Optimum) has a compelling narrative and excellent artwork.

The Urban Farmer

by Kate Burt

Fritz Haeg is an architect and artist who has applied his skills to an original kind of gardening in the inner city. He is the author of *Edible Estates*.

Fritz Haeg isn't perhaps the obvious representative of a revolution in global farming. As an architecture and design academic and practitioner, the American has had his work exhibited at Tate Modern and the Whitney Museum of American Art, and has taught fine art at several US universities. Yet it is last year's community-collaborative project on an inner-city council estate in south London that best showcases his current passion: the urban farm.

Last April, in a discussion about the global food crisis, Gordon Brown announced: 'We need to make great changes in the way we organise food production in the next few years.' High on the list of viable[1] changes is the idea of inner-city agriculture. Which is the theory behind Haeg's concept, detailed in his new book *Edible Estates*: it proposes the replacement of the domestic front lawn in cities with 'an edible landscape'. Last year, to illustrate this point, Haeg was commissioned by the Tate to create a permanent 'edible estate' on a triangle of communal grass in front of a housing estate near Elephant and Castle, bordered on two sides by a main road along which London buses thunder every few minutes.

The aim was to engage and involve the local residents – and together they miraculously transformed a patch of grass previously favoured by dogs and drunks into a luscious agri-plot housing apple and plum trees, a 'forest' of tomato plants, aubergines, squashes, Brussels sprouts, runner beans, sweet peas, a 'salad wing', herbs, edible flowers and 6ft artichoke

---

[1]**viable**   possible, workable

plants. It is also quite beautiful: 'The design was inspired by the ornate, curvy raised flowerbeds you find in front of Buckingham Palace,' explains Haeg. Interestingly, although this space is still accessible by passers-by – unlike the traditional allotment, which Haeg feels is outdated – there has been no theft or vandalism. The London project was mirrored in several locations around the US.

'All the projects I do are rooted in the way that an architect thinks and works,' says Haeg. 'How we live and the spaces we make for ourselves.' And right now, he believes, we need to re-evaluate exactly that, and urgently so – particularly in our over-crowded cities.

As part of its 'One Planet Living' initiative, the World Wildlife Fund calculated our average personal carbon footprint in Britain. Perplexingly, it found that food production and its transport accounts for our greatest use of carbon – 23 per cent per person – beating personal transport, home energy and even shared services (the running of schools, hospitals, banks and so on). These results, combined with food shortages and escalating[2] costs – the price of apples and eggs has risen by 30 per cent in the past year – mean action must be taken, says Haeg. Ornamental urban space is a luxury we can no longer afford, he believes: we need to be growing food on our lawns, greens, driveways and even public parks.

Haeg is not the only one to think it is time for change. The global Consultative Group on International Agricultural Research (CGIAR) strategic alliance estimates that, by 2015, more than half the world's population will be living in urban areas, provoking one of the greatest challenges in the history of agriculture as we try to find a way to keep a lid on food miles and produce enough food for everyone. 'Now, more than ever,' urges Sustain, the alliance for better food and farming, 'we need to grow more food closer to where people live.' And in this climate, it seems that everyone from town planners to head

[2]**escalating**   increasing, worsening

teachers, TV chefs to agri-entrepreneurs[3] are getting excited about farming food in the big smoke.

But is it realistic to turn over our spare urban soil to the cause – and is there really enough of it to do so? Erlk Watson, an urban design director at the town-planning company Turley Associates, strongly believes that inner-city agriculture is the future. As such, he is already advising his clients on ways to incorporate farming into their developments and is particularly excited about the potential for transforming existing space enclosed in the traditionally British city structure, the 'perimeter block' (a row of buildings constructed around an enclosed, private square – typically divided into private gardens). 'Look at an aerial view of London and you'll see there's an enormous amount of private open space contained within these blocks. It is perfect for this urban agricultural revolution,' he says.

Re-apportioning private space might not be as far-fetched as it sounds. Later this month Sustain is hosting a conference, called Growing Food for London, where ideas to be aired include the possibilities of using derelict council facilities, social housing land and unused private gardens for commercial agriculture, as well as the planting of fruit and nut trees in parks and along roads, creating community gardens in public parks and replacing ornamental plants with edible crops. It will also look at alternative food production such as mushroom growing, beekeeping and planting edibles in window boxes, as well as ideas for the little-explored area of rearing livestock in urban areas.

While beekeeping is on the rise in British cities – it is estimated that there are 5,000 beehives in London alone – other urban animal-based edibles are rare. Hunting might be the answer here – squirrel meat has already been seized upon as a sustainable,[4] free-range delicacy in rural Cornwall – could it

---

[3]**entrepreneurs**   people who set up businesses
[4]**sustainable**   using natural resources without destroying the natural
    balance

catch on in cities? Might pigeon pie become a Trafalgar Square speciality; has anyone thought of fox cutlets?

Perhaps more realistic is organised urban livestock rearing. 'There are issues with planning – noise pollution and so on,' says Zeenat Anjani from Sustain, 'but you could definitely raise chickens and other small animals. We hope the Growing Food conference will open more people's minds to these sorts of ideas and get the right people in the same room to talk about what they can do.'

Many are already talking about it. Inspired by the 'victory gardens'[5] of the First and Second World Wars, when civilians were urged to 'dig for victory' to survive the food shortages, Jamie Oliver's newest venture is to inspire the residents of inner-city Rochdale to eat like our wartime forebears and grow their own, while Hugh Fearnley-Whittingstall's new River Cottage series challenges five Bristol familes to transform a derelict patch of land into a fruitful smallholding.

In Middlesbrough, the Groundwork South Tees trust has begun an urban-farming education programme to teach people how to cultivate herbs, vegetables and fruit even if they do not have a garden, by providing containers for patios, balconies and windowsills. There are also sustainable-food grants available to those who want to educate others how to produce their own food in cities, and how to compost effectively to improve typically poor-quality urban soil.

If it comes off, perhaps one of the most high-profile initiatives – still at bid stage – is the Feed the Olympics proposal. It is a radical blueprint from several green organisations outlining how 6,000 acres of land in London could be put to work to grow enough food to provide the 14m-odd meals that will be needed during the 60 days of the 2012 Games, instead of importing it. This would involve creating 2,012 new food-growing spaces across the capital, including community gardens, allotments and roof gardens.

---

[5]**victory gardens**   gardens created to grow food during World War II

Revolutionary? In this country, yes – but we're lagging behind countries such as China, Japan and Cuba, which already have farms integrated into the social, economic and physical structures of their cities; as early as a decade ago Beijing town planners had begun to incorporate agriculture into the urban landscape. The Chinese government also offers courses to aspiring urban farmers and plans to cultivate gardens on nearly 10,000,000 sq ft of roof space over the next 10 years.

Similarly, Argentina's Programa de Agricultura Urbana (PAU) was set up to support city-based farmers in the aftermath of the country's financial collapse. And in Cuba, when the US-led trade embargo[6] resulted in severe food shortages, the government responded by investing in urban farms, providing state-owned plots and teaching relevant skills in schools.

---

[6]**embargo**   a block or bar on trade with another nation or on a particular product

But will it work in Britain? Carole Wright, who manages the communal garden created by Haeg in south London, says it already is. 'It cost less than £5,000 to create and it is capable of feeding three blocks of flats with 24 households each,' she says. 'We run family gardening sessions, Sunday sessions, after-school clubs and also container gardening, so residents can grow things on their balconies too. High-density housing is no barrier – you can grow things out of an old baked-bean can. The more people we can get, the more we can produce. It's not about the size of the land – it's about the maintenance.' She has had no shortage of regular, enthusiastic volunteers – surprisingly most of whom are children.

Wright was delighted when one girl, a moody teenager who described herself as a 'cybergoth', grew her own beet-root. 'You'd never have known she was excited about it,' says Wright, 'but I spotted her one evening with her friends, holding the thing in her hands. "What are you doing with that?" I asked. "Well," she said, "I grew it – I wanted to show my mates." She comes down every day now to water her sun-flowers.'

It's not just about financial and health benefits – Wright has also noticed social benefits. 'People who have not spoken for five years are suddenly chatting again, discussing what they've grown. And it brings together people from different cultures too – they lean over the fence and reminisce about the vegetables they grew in their countries as children – okra, bananas, yams, sweet potatoes.'

Wright describes one gardener, an elderly widow, who has planted an almond tree as a memorial to her late husband and says he would have loved to see how the space had been transformed. 'One guy has even replaced the photo of his family on his mobile phone with a picture of the garden. It's given them so much pride.'

The impact of the garden has been enormous, says Wright. People from further and further away are coming along to get involved, learn new skills and socialise. 'They see it and it's like a lightbulb and they say, "We want our own edible estate." Well, it makes sense, doesn't it?'

Further reading

If you are interested in Fritz Haeg's ideas, read *Edible Estates: Attack on the Front Lawn* (Distributed Art Publishers, 2008), which records the edible estates he has created in California, Kansas and New Jersey, as well as London. There are also contributions from other landscape architects and writers.

The Rough Guide to Climate Change
by Robert Henson

> Robert Henson's *Rough Guide to Climate Change* (Rough Guides Ltd, 2008) gives, as its name suggests, facts, information and opinions on what is happening to our climate and weather patterns, how governments across the world are tackling the problem and what each of us should do about it.

The big picture
Is the planet really warming up?

In a word, yes. Independent teams of scientists have laboriously combed through more than a century's worth of temperature records (in the case of England, closer to 300 years' worth). These analyses all point to a rise of more than 0.7°C (1.3°F) in the average surface air temperature of Earth over the last century.

In recent years global temperatures have spiked dramatically, reaching a new high in 1998. An intense El Niño[1] early that year clearly played a role in the astounding warmth, but things haven't exactly chilled down since then. The first six years of the twenty-first century, along with 1998, were the hottest on record – and quite possibly warmer than any others in the past millennium.

Apart from what temperatures tell us, there's also a wealth of circumstantial evidence to bolster the case that Earth as a whole is warming up.

➤ **Ice** on land and at sea is melting dramatically in many areas outside of interior Antarctica and Greenland. Montana's Glacier National Park is expected to lose its glaciers by 2030. Arctic sea ice has lost nearly half its average summer thickness

---

[1] **El Niño**   a warming of the ocean surface off South America that affects winds and other weather patterns

a

b

Arctic sea ice in 1979 (a) and 2005 (b).

since 1950, and by mid-century the ice may disappear completely each summer, perhaps for the first time in more than a million years. The warmth is already heating up international face-offs over shipping, fishing and oil-drilling rights in parts of the Arctic once written off as inaccessible.

➤ **The growing season** has lengthened across much of the Northern Hemisphere. The most common species of Japan's famed sakura (cherry blossoms) now blooms five days earlier on average in Tokyo than it did fifty years ago. At some higher latitudes, the growing season is now more than two weeks longer than it was in the 1950s – hardly a crisis in itself, but a sign that temperatures are on the increase.

➤ **Mosquitoes, birds and other creatures** are being pushed into new territories, driven to higher altitudes and latitudes by increasing warmth. The range of twelve bird species in Britain shifted north in the 1980s and 1990s by an average of 19km (12 miles). And Inuits[2] in the Canadian Arctic report the arrival over the last few years of barn swallows, robins, black flies and other previously unseen species.

But don't many experts claim that the science is uncertain?
There is plenty of uncertainty about details in the global-warming picture: exactly how much it will warm, the locations where rainfall will increase or decrease, and so forth. Some of this uncertainty is due to the complexity of the processes involved, and some of it is simply because we don't know how individuals, corporations and governments will change their greenhouse emissions over time. But there's near-unanimous agreement that global climate is already changing and that fossil fuels are at least partly to blame.

The uncertainty that does exist has been played both ways in the political realm. Sceptics use it to argue for postponing action, while others point out that many facets of life require acting in the face of uncertainty (buying insurance against health or fire risks, for example).

Is a small temperature rise such a big deal?
While a degree or so of warming may not sound like such a big deal, the rise has been steeper in certain locations, including the

---

[2]**Inuits**   people of the Arctic regions

Arctic, where small changes can become amplified into bigger ones. The warming also serves as a base from which heat waves become that much worse – especially in big cities, where the heat-island effect[3] comes into play. Like a thermodynamic echo chamber, the concrete canyons and oceans of pavement in a large urban area heat up more readily than a field or forest, and they keep cities warmer at night. During the most intense hot spells of summer, cities can be downright deadly, as evidenced by the hundreds who perished in Chicago in 1995 and the thousands who died in Paris in 2003.

How could humans change the whole world's climate?

By adding enormous quantities of carbon dioxide and other greenhouse gases[4] to the atmosphere over the last 150 years. As their name implies, these gases warm the atmosphere, though not literally in the same way a greenhouse does. The gases absorb heat that's radiated by Earth, but they release only part of that heat to space, which results in a warmer atmosphere.

The amount of greenhouse gas we add is staggering – in carbon dioxide alone, the total is more than thirty billion metric tonnes per year, which is more than four metric tonnes per person per year. And that gas goes into an atmosphere that's remarkably shallow. If you picture Earth as a soccer ball, the bulk of the atmosphere would be no thicker than a sheet of paper wrapped around that ball.

Even with these facts in mind, there's something inherently astounding about the idea that a few gases in the air could wreak havoc around the world. However, consider this: the eruption of a single major volcano – such as Krakatoa in 1883 – can throw enough material into the atmosphere to cool global climate by more than 1°C (1.8°F) for over a year. From that perspective, it's not so hard to understand how the millions of

---

[3]**heat-island effect**   an urban area that is hotter than surrounding rural areas
[4]**greenhouse gases**   gases such as carbon dioxide and methane that contribute to the greenhouse effect

engines and furnaces spewing out greenhouse gases each day across the planet, year after year, could have a significant effect on climate. (If automobiles spat out chunks of charcoal every few blocks in proportion to the invisible carbon dioxide they emit, the impact would be more obvious.) Yet many people respond to the threat of global warming with an intuitive, almost instinctive denial.

How do the rainforests fit into the picture?

The destruction of rainforests across the tropics is a significant contributor to climate change, accounting for roughly a fifth of recent human-produced CO_2 emissions. Tropical forests hold nearly half of the carbon present in vegetation around the world. When they're burned to clear land, the trees, soils and undergrowth release CO_2. Even if the land is eventually abandoned, allowing the forest to regrow, it would take decades for nature to reconcile the balance sheet through the growth of replacement trees that pull carbon dioxide out of the air. In addition to the CO_2 from the fires, bacteria in the newly exposed soil may release more than twice the usual amount of another greenhouse gas, nitrous oxide, for at least two years. Brazil's National Institute for Amazon Research estimates that deforestation puts four times more carbon into the atmosphere than the nation's fossil-fuel burning does.

Rainforests also cool the climate on a more local level, their canopy helping to trap moisture and allow it to slowly evaporate, providing a natural air-conditioning effect. When the rainforest has been slashed and burned over large areas, hotter and dryer conditions often set in, although the exact strength of this relationship is difficult to quantify. Across eastern Brazil, where nearly 20% of the Amazonian rainforest has been destroyed, 2005 saw the region's worst drought in a century, perhaps related to changes in the nearby Atlantic and to rain-suppressing smoke from fires as well as to the deforestation itself.

By contrast, in mid-latitude and polar regions, forests actually tend to warm the climate.

The outlook
How hot will it get?

According to the 2007 IPCC[5] report, the global average temperature is likely to rise anywhere from 1.1°C to 6.4°C (2.0–11.5°F) by 2080–2099, relative to 1980–1999. This range reflects uncertainty about the quantities of greenhouse gases we'll add to the atmosphere in coming decades and also about how the global system will respond to those gases. Some parts of the planet, such as higher latitudes, will heat up more than others. The warming will also lead to a host of other concerns – from intensified rains to melting ice – that are liable to cause more havoc than the temperature rise in itself.

Is global warming necessarily a bad thing?

Whether climate change is bad, good or neutral depends on your perspective. Some regions – and some species – may benefit, but many more will suffer intense problems and upheavals. And some of the potential impacts, such as a major sea-level rise, increased flooding and droughts, more major hurricanes and many species being consigned to extinction, are bad news from almost any perspective. So while it may be a bit of a reach to think in terms of 'saving the planet' from global warming, it's perfectly valid to think about preserving a climate that's sustaining to as many of Earth's residents as possible.

Perhaps the more pertinent question is whether the people and institutions responsible for producing greenhouse gases will bear the impacts of their choices, or whether others will – including those who had no say in the matter. Indeed, people in the poorest parts of the world – such as Africa – will generally be least equipped to deal with climate change, even if the changes are no worse there than elsewhere. Yet those regions have released only a small fraction of the gases that are causing the changes.

---

[5]**IPCC**   Intergovernmental Panel on Climate Change

Will anyone be killed or displaced?

Quantifying the human cost of climate change is exceedingly difficult. Weather-related disasters kill thousands of people each year, regardless of long-term changes in the climate. Many of the projected impacts of global warming on society are the combined effects of climate change and population growth (some claim the latter is far more important than the former). For this reason, it's hard to separate out how much of the potential human suffering is due to each factor.

In the decades to come, the warming of the planet and the resulting rise in sea level will likely begin to force people away from some coastlines. Low-lying islands are already vulnerable, and entire cities could eventually be at risk. The implications are especially sobering for countries such as Bangladesh, where millions of people live on land that may be inundated before the century is out.

Another concern is moisture – both too much and too little. In many areas rain appears to be falling in shorter but heavier deluges conducive to flooding. However, drought also seems to be becoming more prevalent. Changes in the timing of rainfall and runoff could complicate efforts to ensure clean water for growing populations, especially in the developing world.

Warming temperatures may also facilitate the spread of vector-borne diseases[6] such as malaria and dengue fever. The World Health Organization estimates that in 2000 alone, more than 150,000 people died as a result of direct and indirect climate-change impacts.

Will agriculture suffer?

That depends on where the farming and ranching is done. Global agricultural productivity is predicted to go *up* over the next century, thanks to the extra CO_2 in the atmosphere and now-barren regions becoming warm enough to bear crops. However, the rich world looks set to reap the benefits: crop

---

[6]**vector-borne diseases**   diseases carried by organisms such as the mosquito

yields in the tropics, home to hundreds of millions of subsistence farmers, are likely to drop.

And wildlife?

Because climate is expected to change quite rapidly from an evolutionary point of view, we can expect major shocks to some ecosystems – especially in the Arctic – and possibly a wholesale loss of species. According to a 2004 study led by Chris Thomas of the University of Leeds and published in the journal *Nature*, climate change between now and 2050 may commit as many as 37% of all species to eventual extinction – a greater impact than that from global habitat loss due to human land use. Similar figures emerged from the 2007 IPCC report, which pegs the percentage of plant and animal species that are at risk from a temperature rise of 1.5–2.5°C (2.7–4.5°F) at 20–30%.

What can we do about it?

Which countries are emitting the most greenhouse gases?

For many years the United States was in first place, with 30% of all of the human-produced greenhouse emissions to date and about 20% of the current yearly totals – despite having only a 5% share of global population. However, China is now taking the lead. Its emissions are *much* lower per capita, but due to its growing population and affluence, China will overtake the US as the world's leading greenhouse emitter by 2008.[7] The world's industrialized countries vary widely in how much they have increased or decreased their total emissions since 1990. Some of the decreases were due to efficiency gains, while others were due to struggling economies.

Does the growth of China and India make a solution impossible?

Not necessarily. Although its growth in its coal production is hugely worrisome, China is already making progress on vehicle fuel efficiency and other key standards. And because so much of the development in China and India is yet to come, there's a

---

[7]China has now overtaken the US as the world's largest producer of greenhouse gases.

window of opportunity for those nations to adopt the most efficient technologies possible. At the same time, the sheer numbers in population and economic growth for these two countries are daunting indeed – all the more reason for prompt international collaboration on technology sharing.

If oil runs out, does that solve the problem?

Hardly. It's true that if oil resources do 'peak' in the next few years, as some experts believe, we're likely to see economic downswings, and those could reduce oil-related emissions, at first over periods of a few years and eventually for good. The same applies to natural gas, although that peak could arrive decades later. Then, the question becomes what fuel sources the world will turn to: coal, nuclear, renewables or some combination of the three. If the big winner is coal – or some other, less-proven fossil source such as shale or methane hydrates[8] – it raises the potential for global warming far beyond anything in current projections.

Even if renewables win the day later in this century, we're still left with the emissions from today's stocks of oil, gas and coal, many of which would likely get burned between now and that eco-friendly transition. With this in mind, research has intensified on sequestration – how carbon might be safely stored underground. The idea appears promising, but big questions remain.

Won't nature take care of global warming in the long run?

Only in the *very* long run. The human enhancements to the greenhouse effect could last the better part of this millennium. Assuming that it takes a century or more for humanity to burn through whatever fossil fuels it's destined to emit, it will take hundreds more years for those greenhouse gases to be absorbed by Earth's oceans.

There are few analogies in the geological past for such a drastic change in global climate over such a short period, so it's

---

[8]**methane hydrates**   gas found in ice deep in the oceans

impossible to know what will happen after the human-induced greenhouse effect wanes. All else being equal, cyclical changes in Earth's orbit around the Sun can be expected to trigger an ice age sometime within the next 50,000 years, and other warmings and coolings are sure to follow. In the meantime, we'll have our hands full dealing with the next century and the serious climate changes that our way of life may help bring about.

Further reading

The Rough Guide to Climate Change is written in a clear, level-headed style and is well worth reading. It has an introduction by the famous environmentalist and scientist James Lovelock. In his books, *Gaia* (Gaia Books Ltd, 2005) and *The Revenge of Gaia* (Penguin Books, 2007), he sees Earth as a self-regulating planet that will sustain life, although individual species, such as ourselves, may die out and others take their place.

How Many Light Bulbs Does it Take to Change a Planet?

by Tony Juniper

In his book Tony Juniper sets out ninety-five points suggesting ways to limit the effects of climate change. He discusses reducing toxic emissions, protecting rainforests, and promoting appropriate farming methods, along with many other related issues.

In 1990 I had the remarkable and rare experience of virtually witnessing the extinction of a wild species. I was a member of a research team sent to northeast Brazil by Bird Life International with the task of determining the wild status of one of the world's rarest birds. This particular species, the Spix's macaw, was known to be very rare, but not as rare as we found it to be. After an intensive period of fieldwork and a thorough examination of all the published and museum evidence we could find, we reached the unavoidable conclusion that the single male we had found was the last of his kind still living in the woodlands that comprised the bird's natural habitat. With no partner and, therefore, no means to continue its kind, Spix's macaw was effectively extinct in the wild.

The species had been brought to this desperate state of affairs because of a centuries-long process of habitat degradation.[1] The birds' native habitat, the dry *caatinga*[2] thorn forest, was initially reduced by burning and grazing, and then by large-scale irrigated agriculture producing sugar and soya for the world's commodity[3] markets. On top of this, the Spix's macaw was prized by the collectors of rare parrots. The rarer it got, the more they wanted it. In the end the last few (except one) were

---

[1] **degradation**   ruin
[2] **caatinga**   thorn forest growing where there is little rainfall
[3] **commodity**   something (such as an agricultural product) that can be processed and sold on

Believed to be extinct in the wild.

captured and sold on to collectors, including some in Europe. At the end, the birds were changing hands for tens of thousands of dollars each. The trapping was driven by demand on the other side of the world, and nearly all the money was made by middle men in the big cities in the south of Brazil. As if this wasn't bad enough, there is a suspicion that the last nesting birds were attacked by introduced African bees. These insects are very aggressive when seeking out new locations for hives and may have come into conflict with nesting parrots – whose holes make ideal locations for the bees to set up home.

The last wild bird we found in 1990 died in 2000, leaving the Spix's macaw with the official status of 'extinct in the wild'. There were, however, some 13 of the birds in captivity. These formed the nucleus of a breeding programme that has today (despite many serious setbacks and complications) succeeded in establishing a population of about 100 in captivity. I hope one day that a reintroduction will take place – assuming that any of the birds' rare woodland habitat is left for them to be released back in to. They should of course never have been trapped in the first place, and it would be far better if there was instead an active programme to restore the bird's numbers in the wild state.

What happened to the wild population of Spix's macaws was for me just one dramatic example of what is going on right across our planet.

Solution: Embark on the large-scale restoration of certain types of habitat, otherwise many currently protected areas will be unable to sustain their unique biodiversity into the long term

From a biodiversity point of view, the place where I live is one of the most ruined parts of the world. It is a pleasant place and people want to live there, but the native wildlife is massively depleted. I live in Cambridge, a city at the edge of the fens of eastern England.

The fens were once a vast wetland stretching from the edge of the city north to the Wash and into Norfolk and Lincolnshire. A mosaic of reeds and sedges, wet grasslands and scrubby woodlands dotted with ponds and lakes, the fens brimmed with birds, fish and other wildlife. Today, this rich lowland wetland is nearly all drained, and is one of the most intensively farmed parts of Europe.

The Romans first saw the agricultural potential of the fens and embarked on a programme to drain the rich peaty soils so as to grow crops. Attempts to claim the land for farmland continued during medieval times, but it was not until the 1630s that contractors working for King Charles I began large scale drainage in earnest.

The leader of one syndicate of landowners was the Earl of Bedford. He employed a Dutch drainage engineer called Cornelius Vermuyden, who masterminded the near total drainage of the fens. In the face of spirited opposition from locals, who were set to lose their livelihoods – fishing and wildfowling – for the benefit of already wealthy landowners, huge new drainage 'rivers' were excavated. The ancient fenland habitat was mostly destroyed, and in its place came increasingly intensive arable farming.

Surprisingly perhaps, more than three hundred years after Vermuyden cut his long, straight drainage ditches, I find powerful inspiration in this landscape. This is not born from the tragedy of what has been lost, it comes from the rebirth of the landscape and the return of its wildlife. What were until recently intensively farmed wheat fields are now being restored as fenland, and the effect is amazing.

Only a few tiny fragments of original fenland escaped the destruction: in total less than one-third of 1 per cent of the original area survived. Wicken Fen is one of the little corners to have been spared. Even when Charles Darwin was studying at Cambridge in the 1820s it was known to be an important place. Before leaving England on his famous voyage of evolutionary discovery aboard the *Beagle*, he regularly went to Wicken to collect insects.

Considering this history it is perhaps fitting that Wicken Fen became one of England's first official nature reserves. The National Trust bought the initial little strip in 1899 for just £10. From then on, bits of surviving fen have been added until a good chunk of native habitat sat protected like an island in a great hostile sea of sugar beet, wheat, barley and oil seed rape. The remaining patch of fen, puffed up with water, sits above the land around it, which has shrunk back as drainage ditches suck away the life-giving water.

In its isolated state it was clear that the wildlife of Wicken Fen would decline. Species would be lost and in the longer term the richness of the reserve would steadily diminish. To avoid this, Wicken needed to be expanded so that it could function more naturally, and thus hang on to its full complement of wild species. This has now begun in earnest, and for me it is hugely symbolic. The restoration of fenland signals the shift from the defensive protection of the last remnants of important habitat to a proactive campaign of large-scale reclamation[4] and recovery. Other large-scale restoration schemes are underway in the

---

[4]**reclamation**   restoration

intensively farmed east of England, at Woodwalton Fen and Lakenheath, for example.

Even some of the damage I witnessed being inflicted on wildlife havens during my own youthful forays around Oxfordshire is being reversed. Otmoor was a large expanse of wet grassland to the north of Oxford that I saw being progressively drained during the 1970s, to the point where only a few tiny bits of the original habitat remained. Today, however, the Royal Society for the Protection of Birds has bought up sizable chunks, and in a few years it has restored much of Otmoor to its former glory. Even red kites can be seen there today, following the successful reintroduction of these birds on to the nearby Chiltern Hills during the 1990s.

Wicken Fen is now fast expanding as the National Trust buys up neighbouring parcels of farmland whenever they become available. The long-term vision of the National Trust is to create a restored wetland habitat all the way to Cambridge – more than 16 km (10 miles) to the south.

Wicken Fen restores more than nature, it replenishes energy and optimism. Even in the melancholy of mid-winter it is uplifting to look south from the edge of the old fen over the new lands recently reclaimed from ploughs and pesticides. Under a vast sky, the brown land is once again alive. Native roe deer and hares cross the old drainage ditches, now clogged with beige reed stems. Hen harriers and short-eared owls quarter the land in search of prey. In the cool winter of 2005–06, a rare great grey shrike visited from the continent and stayed for some months. I watched it several times with my children.

Solution: Offer support to those societies around the world who are still living sustainably

Many societies live closely with the untouched and diverse ecosystems[5] that have supported them for many centuries, in

---

[5]**ecosystem**   a system in which organisms depend on each other within their environment/habitat

some cases for millennia. By working with such societies, supporting them in their own efforts to protect their culture and way of life, it will be possible to secure the protection of a large proportion of the Earth's biodiversity.

While the people of the fens lost their campaign to retain the landscape that sustained them nearly four centuries ago, battles to save the local environment are still being fought by communities around the world. From the Norwegian Arctic to the tropical rainforests of Colombia, people are engaged in struggles to protect ways of life that have existed in sustainable coexistence with nature for generations.

Such sustainable societies are now mainly in remoter regions. These areas, previously isolated from outside influences, are now under pressure. New roads are penetrating once inaccessible territory as demand for natural resources such as oil, gas, wood and metals takes companies and government agencies into the last relatively untouched parts of our globe.

Several environmental organizations are assisting local peoples to gain more control over land that they have occupied in some cases for thousands of years. They are offering this assistance as a conservation strategy. And it works.

If one reviews a map of deforestation across the southern half of the Brazilian Amazon, and sees the consequences of logging, mining, charcoal production and the clearance of land for cattle and soya production, then one is confronted with ecological damage on a massive scale. Very importantly, the areas of remaining intact forest coincide very closely with the lands controlled by the traditional indigenous inhabitants. Their culture and lifestyles are far more conducive[6] to sustaining the forests than the extractive depredations[7] adopted by the recent in-comers. The latter are only interested in the lucrative profits of mahogany logging, mining and soya farming.

---

[6]**conducive**   encouraging
[7]**depredations**   damage

Indigenous people[8] account for some 80 to 90 per cent of human cultural diversity and their lands include territory located in some 80 per cent of the Earth's major habitat types. Because so many of the areas occupied by indigenous peoples are in the diverse tropical systems such as rainforests, mangroves and in-shore reefs, it is possible that more than 80 per cent of the Earth's species are in lands where indigenous societies still live. If it is the local people who have rights over the land, rather than the transnational corporations, then the possibilities for conserving biodiversity can be considerably enhanced.

This can be seen in Papua New Guinea, for example, where customary land rights have greater recognition and protection than is the case with many other forest peoples around the world. As a consequence Papua New Guinea – one of the most culturally and biologically diverse places on Earth – has so far suffered relatively light levels of environmental damage compared to that inflicted elsewhere at the hands of loggers, miners and oil drillers.

The Rainforest Foundation is one organization that is helping communities resist such destruction, by providing them with the technical capacity to make legal claims over the land they occupy. Lacking the resources of government departments and international companies, the forest communities would not stand much chance without this kind of support. Even with modest resources a big difference can thus be made. Over the last fifteen years, the Rainforest Foundation estimates that by such methods it has secured nearly 100,000 square kilometres of forest (nearly 400,000 square miles – roughly the size of Scotland and Wales combined).

Solution: Protect what remains of natural forest around the world

So called 'old-growth' forests now cover only about 10 per cent of the Earth's land surface, but may contain as much as 70 per cent

---

[8]**indigenous people**   people of a region at the time of contact with the outside world

of the world's plant and animal species. These forests are also important in helping to maintain climatic stability, not least because they hold vast quantities of carbon. Protecting these forests is thus a priority, and we need to act to ensure their long-term survival, for example by cracking down on illegal logging, establishing new protected areas, and changing consumer behaviour. It is also essential to include incentives for the conservation of old-growth forests in future international climate-change agreements.

Old-growth forests are amazing ecosystems. In late 1996 I visited the ancient old-growth forests of northern Finland. I was there to discuss strategy with campaigners working to save the taiga, the boreal forests that ring the Earth across northern North America, Scandinavia and Siberia. The taiga is under pressure everywhere. The main threat is rapacious[9] logging, often using the practice of clear-cutting – which for a complex ecosystem is the equivalent of blanket bombing a city. Boreal timber – predominantly softwoods from conifers – is used for construction and joinery, and vast quantities of it are used to make paper.

Old-growth forests are those that have not been logged, at least for some hundreds of years. In northern Scandinavia these forests mainly consist of spruce and pine, with some birch and aspen. In the summer they are alive with birds and insects and contain many unique plants, including lichens and mosses that only live in the un-logged old growth. Late in the year, when the sun slides only briefly across the southern horizon, the forests fall silent. Through the long winter months, locked in frost and snow, they take on a quiet, haunting beauty.

One day during my 1996 trip I sat alone in an area of old-growth forest on the border between Russia and Finland. There was only a brief period of daylight. During the early dusk I had a rare view of one of the few creatures able to tough it out there for the whole winter: a great grey owl. This is one of the few

---

[9]**rapacious**   greedy

birds of prey that can actually hear its prey moving *beneath* the snow, and thus can survive on mice and lemmings all through the northern winter.

Even in Scandinavia, an environmentally aware and heavily forested part of the world, only a few per cent of the original forest remains un-logged. The rest has either been clear-cut, or trees have been selectively felled. Although the clear cuts are mostly replanted or allowed to regenerate, and while some of those areas that are logged are increasingly well managed, the importance for biodiversity of the old-growth natural forest is unique. Some species, such as white-backed woodpeckers and European flying squirrels, simply will not occupy anything other than old growth, and what is left needs to be protected.

In Scandinavia the protection of old-growth forest might be expected to be a relatively straightforward matter, but in many developing countries the situation could hardly be more challenging. Massive international debts, corrupt public institutions and fast-growing populations demanding more land and resources means that many countries are fighting losing battles against the loss of old-growth forest. Add to this set of pressures the impacts of excessive consumption[10] in the richer countries and it is clear that there is an enormous momentum driving deforestation.

A case in point is the deforestation taking place on the Indonesian island of Sumatra. Companies such as Asia Pulp and Paper have been responsible for the devastation of huge swathes of old-growth forest so as to make paper that people largely use only once and then condemn to a landfill or incinerator.

As far as paper is concerned it is essential to achieve higher recycling rates. The less virgin paper we use and the more waste is recycled back into a useable product, the less will be the consumer 'pull' that is damaging forests everywhere. Protected area networks need to embrace the last old-growth forests, especially

---

[10]**consumption**   using up

those with the highest biodiversity and the largest numbers of unique species.

Further reading

The book *How Many Light Bulbs Does it Take to Change a Planet?* (Quercus, 2007), from which the extract comes, gives a very comprehensive account of climate change, and looks at contributing factors such as travel, resources and food and how we can bring about change in our lifestyles.

The Final Straw

by Simon Armitage

This impressive poem comes from Simon Armitage's collection
Tyrannosaurus Versus the Corduroy Kid (Knopf Publishing Group,
2008). The Spix's macaw is now extinct in the wild.

Corn, like the tide coming in. Year on year,
fat, flowing grain, as it had always grown.
We harvested clockwise, spiralling home
over undulations of common land
till nothing remained but a hub of stalks
where the spirit of life was said to lurk.

So childless couples were offered the scythe –
the men invited to pocket the seed,
the women to plait dolls from the last sheaf.

But a Spix's macaw flapped from the blade,
that singular bird of the new world, one
of a kind. A rare sight. And a sign, being
tail-feathers tapering out of view, being

blueness lost in the sun, being gone.

Further reading

If the work of Simon Armitage appeals to you, you might like to read
his *Selected Poems* (Faber and Faber, 2001). There you can find a wide
range of his work from his different volumes of poetry. Many of his
poems are studied at GCSE, such as *Kid*, written from the perspective
of Robin, friend of Batman the comic strip hero.

The Forgotten Enemy

by Arthur C. Clarke

Arthur C. Clarke was one of the most prolific and highly acclaimed science fiction writers of his generation. In this story, the aging Professor Millward finds himself marooned in an environment that has undergone a vast change.

The thick furs thudded softly to the ground as Professor Millward jerked himself upright on the narrow bed. This time, he was sure, it had been no dream; the freezing air that rasped against his lungs still seemed to echo with the sound that had come crashing out of the night.

He gathered the furs around his shoulders and listened intently. All was quiet again: from the narrow windows in the western walls long shafts of moonlight played upon the endless rows of books, as they played upon the dead city beneath. The world was utterly still; even in the old days the city would have been silent on such a night, and it was doubly silent now.

With weary resolution Professor Millward shuffled out of bed, and doled a few lumps of coke[1] into the glowing brazier. Then he made his way slowly toward the nearest window, pausing now and then to rest his hand lovingly on the volumes he had guarded all these years.

He shielded his eyes from the brilliant moonlight and peered out into the night. The sky was cloudless: the sound he had heard had not been thunder, whatever it might have been. It had come from the north, and even as he waited it came again.

Distance had softened it, distance and the bulk of the hills that lay beyond London. It did not race across the sky with the wantonness[2] of thunder, but seemed to come from a single

---

[1]**coke**   a residue of coal, used for fuel
[2]**wantonness**   shamelessness

point far to the north. It was like no natural sound that he had ever heard, and for a moment he dared to hope again.

Only Man, he was sure, could have made such a sound. Perhaps the dream that had kept him here among these treasures of civilisation for more than twenty years would soon be a dream no longer. Men were returning to England, blasting their way through the ice and snow with the weapons that science had given them before the coming of the Dust. It was strange that they should come by land, and from the north, but he thrust aside any thoughts that would quench the newly kindled flame of hope.

Three hundred feet below, the broken sea of snowcovered roofs lay bathed in the bitter moonlight. Miles away the tall stacks of Battersea Power Station glimmered like thin white ghosts against the night sky. Now that the dome of St Paul's had collapsed beneath the weight of snow, they alone challenged his supremacy.

Professor Millward walked slowly back along the bookshelves, thinking over the plan that had formed in his mind. Twenty years ago he had watched the last helicopters climbing heavily out of Regent's Park, the rotors churning the ceaselessly falling snow. Even then, when the silence had closed around him, he could not bring himself to believe that the North had been abandoned forever. Yet already he had waited a whole generation, among the books to which he had dedicated his life.

In those early days he had sometimes heard, over the radio which was his only contact with the South, of the struggle to colonise the now temperate lands of the Equator. He did not know the outcome of that far off battle, fought with desperate skill in the dying jungles and across deserts that had already felt the first touch of snow. Perhaps it had failed; the radio had been silent now for fifteen years or more. Yet if men and machines were indeed returning from the north – of all directions – he might again be able to hear their voices as they spoke to one another and to the lands from which they had come.

Professor Millward left the University building perhaps a dozen times a year, and then only through sheer necessity. Over the past two decades he had collected everything he needed from the shops in the Bloomsbury area, for in the final exodus vast supplies of stocks had been left behind through lack of transport. In many ways, indeed, his life could be called luxurious: no professor of English literature had ever been clothed in such garments as those he had taken from an Oxford Street furrier's.

The sun was blazing from a cloudless sky as he shouldered his pack and unlocked the massive gates. Even ten years ago packs of starving dogs had hunted in this area, and though he had seen none for years he was still cautious and always carried a revolver when he went into the open.

The sunlight was so brilliant that the reflected glare hurt his eyes; but it was almost wholly lacking in heat. Although the belt of cosmic dust through which the Solar System was now passing had made little visible difference to the sun's brightness, it had robbed it of all strength. No one knew whether the world would swim out into the warmth again in ten or a thousand years, and civilisation had fled southward in search of lands where the word 'summer' was not an empty mockery.

The latest drifts had packed hard and Professor Millward had little difficulty in making the journey to Tottenham Court Road. Sometimes it had taken him hours of floundering through the snow, and one year he had been sealed in his great concrete watchtower for nine months.

He kept away from the houses with their dangerous burdens of snow and their Damoclean icicles, and went north until he came to the shop he was seeking. The words above the shattered windows were still bright: 'Jenkins & Sons. Radio and Electrical. Television A Specialty.'

Some snow had drifted through a broken section of roofing, but the little upstairs room had not altered since his last visit a dozen years ago. The all-wave radio still stood on the table, and empty tins scattered on the floor spoke mutely of the

_ lonely hours he had spent here before all hope had died. He wondered if he must go through the same ordeal again.

Professor Millward brushed the snow from the copy of *The Amateur Radio Handbook for 1965,* which had taught him what little he knew about wireless. The testmeters and batteries were still lying in their half-remembered places, and to his relief some of the batteries still held their charge. He searched through the stock until he had built up the necessary power supplies, and checked the radio as well as he could. Then he was ready.

It was a pity that he could never send the manufacturers the testimonial they deserved. The faint 'hiss' from the speaker brought back memories of the BBC, of the nine o'clock news and symphony concerts, of all the things he had taken for granted in a world that was gone like a dream. With scarcely controlled impatience he ran across the wave-bands, but everywhere there was nothing save that omnipresent hiss. That was disappointing, but no more: he remembered that the real test would come at night. In the meantime he would forage among the surrounding shops for anything that might be useful.

It was dusk when he returned to the little room. A hundred miles above his head, tenuous and invisible, the Heaviside Layer[3] would be expanding outward toward the stars as the sun went down. So it had done every evening for millions of years, and for half a century only, Man had used it for his own purposes, to reflect around the world his messages of hate or peace, to echo with trivialities or to sound with music once called immortal.

Slowly, with infinite patience, Professor Millward began to traverse the shortwave bands that a generation ago had been a babel of shouting voices and stabbing morse. Even as he listened, the faint hope he had dared to cherish began to fade within him. The city itself was no more silent than the once-crowded oceans

---

[3]**Heaviside Layer**   a layer of gas surrounding the Earth that can reflect radio waves

of ether. Only the faint crackle of thunderstorms half the world away broke the intolerable stillness. Man had abandoned his latest conquest.

Soon after midnight the batteries faded out. Professor Millward did not have the heart to search for more, but curled up in his furs and fell into a troubled sleep. He got what consolation he could from the thought that if he had not proved his theory, he had not disproved it either.

The heatless sunlight was flooding the lonely white road when he began the homeward journey. He was very tired, for he had slept little and his sleep had been broken by the recurring fantasy of rescue.

The silence was suddenly broken by the distant thunder that came rolling over the white roofs. It came – there could be no doubt now – from beyond the northern hills that had once been London's playground. From the buildings on either side little avalanches of snow went swishing out into the wide street; then the silence returned.

Professor Millward stood motionless, weighing, considering, analysing. The sound had been too long-drawn to be an ordinary explosion – he was dreaming again – it was nothing less than the distant thunder of an atomic bomb, burning and blasting away the snow a million tons at a time. His hopes revived, and the disappointments of the night began to fade.

That momentary pause almost cost him his life. Out of a side-street something huge and white moved suddenly into his field of vision. For a moment his mind refused to accept the reality of what he saw; then the paralysis left him and he fumbled desperately for his futile revolver. Padding toward him across the snow, swinging its head from side to side with a hypnotic, serpentine motion, was a huge polar bear.

He dropped his belongings and ran, floundering over the snow toward the nearest buildings. Providentially[4] the

---

[4]**providentially**   luckily

Underground entrance was only fifty feet away. The steel grille was closed, but he remembered breaking the lock many years ago. The temptation to look back was almost intolerable, for he could hear nothing to tell how near his pursuer was. For one frightful moment the iron lattice resisted his numbed fingers. Then it yielded reluctantly and he forced his way through the narrow opening.

Out of his childhood there came a sudden, incongruous[5] memory of an albino ferret he had once seen weaving its body ceaselessly across the wire netting of its cage. There was the same reptile grace in the monstrous shape almost twice as high as a man, that reared itself in baffled fury against the grille. The metal bowed but did not yield beneath the pressure; then the bear dropped to the ground, grunted softly and padded away. It slashed once or twice at the fallen haversack, scattering a few

---

[5]**incongruous**   odd, inconsistent

tins of food into the snow, and vanished as silently as it had come.

A very shaken Professor Millward reached the University three hours later, after moving in short bounds from one refuge to the next. After all these years he was no longer alone in the city. He wondered if there were other visitors, and that same night he knew the answer. Just before dawn he heard, quite distinctly, the cry of a wolf from somewhere in the direction of Hyde Park.

By the end of the week he knew that the animals of the North were on the move. Once he saw a reindeer running southward, pursued by a pack of silent wolves, and sometimes in the night there were sounds of deadly conflict. He was amazed that so much life still existed in the white wilderness between London and the Pole. Now something was driving it southward, and the knowledge brought him a mounting excitement. He did not believe that these fierce survivors would flee from anything save Man.

The strain of waiting was beginning to affect Professor Millward's mind, and for hours he would sit in the cold sunlight, his furs wrapped around him, dreaming of rescue and thinking of the way in which men might be returning to England. Perhaps an expedition had come from North America across the Atlantic ice. It might have been years upon its way. But why had it come so far north? His favourite theory was that the Atlantic ice-packs were not safe enough for heavy traffic further to the south.

One thing, however, he could not explain to his satisfaction. There had been no air reconnaissance;[6] it was hard to believe that the art of flight had been lost so soon.

Sometimes he would walk along the ranks of books, whispering now and then to a well-loved volume. There were books here that he had not dared to open for years, they reminded him so poignantly of the past. But now as the days grew

---

[6]**reconnaissance**   investigation, exploration

longer and brighter, he would some times take down a volume
of poetry and re-read his old favourites. Then he would go to
the tall windows and shout the magic words over the
rooftops, as if they would break the spell that had gripped the
world.

It was warmer now, as if the ghosts of lost summers had
returned to haunt the land. For whole days the temperature
rose above freezing, while in many places flowers were breaking
through the snow. Whatever was approaching from the north
was nearer, and several times a day that enigmatic[7] roar would
go thundering over the city, sending the snow sliding upon a
thousand roofs. There were strange, grinding undertones that
Professor Millward found baffling and even ominous. At times
it was almost as if he were listening to the clash of mighty
armies, and sometimes a mad but dreadful thought came into
his mind and would not be dismissed. Often he would wake in
the night and imagine he heard the sound of mountains moving to the sea.

So the summer wore away, and as the sound of that distant
battle drew steadily nearer Professor Millward was the prey of
ever more violently alternating hopes and fears. Although he
saw no more wolves or bears – they seemed to have fled southward – he did not risk leaving the safety of his fortress. Every
morning he would climb to the highest window of the tower
and search the northern horizon with field-glasses. But all he
ever saw was the stubborn retreat of the snows above
Hampstead, as they fought their bitter rearguard action against
the sun.

His vigil ended with the last days of the brief summer. The
grinding thunder in the night had been nearer than ever
before, but there was still nothing to hint at its real distance
from the city. Professor Millward felt no premonition as he
climbed to the narrow window and raised his binoculars to the
northern sky.

---

[7]**enigmatic**   puzzling, mysterious

As a watcher from the walls of some threatened fortress might have seen the first sunlight glinting on the spears of an advancing army, so in that moment Professor Millward knew the truth. The air was crystal-clear, and the hills were sharp and brilliant against the cold blue of the sky. They had lost almost all their snow. Once he would have rejoiced at that, but it meant nothing now.

Overnight, the enemy he had forgotten had conquered the last defences and was preparing for the final onslaught. As he saw that deadly glitter along the crest of the doomed hills, Professor Millward understood at last the sound he had heard advancing for so many months. It was little wonder he had dreamed of mountains on the march.

Out of the North, their ancient home, returning in triumph to the lands they had once possessed, the glaciers had come again.

Further reading

If you enjoyed *The Forgotten Enemy*, you may also enjoy the author's other short stories. Most of them can be found in *The Collected Stories of Arthur C. Clarke* (Gollancz, 2001). There is also an audio edition (Fantastic Audio, 2001).

Activities

Green Boy

Before you read

1   Many fantasy novels and films have a message about how we should treat each other or our world. What messages did you pick up from what you have read or seen? Exchange your ideas with a partner.

What's it about?

2   Work with a partner to discuss the forest. How is it described in the first few lines? Now consider the strange things about it. Make a list of some of them, starting with:

no insects to be seen, no birds

3   Lou cannot speak. He is regarded as having special powers. What is the first sign that this might be true (see pages 151–152)? Make some notes with a partner.

Thinking about the text

4   Working in a small group, act out the text as a drama. Decide:
   - what characters to include besides Trey and Lou
   - how you will portray Lou, who does not speak
   - whether you need any simple props (for example, to represent the pit)
   - how you will convey that there is a forest around you
   - how you will convey fear and panic.

5   In many ways Pangaia is like our own world. Record a list of similarities. Then consider how it is different. Share your list with a partner and discuss your ideas. Now refer back to question 1. What message do you think *Green Boy* has about our own world? Write a paragraph to explain.

6   The author describes the scene in which the children come across the millipedes vividly, sometimes using compound words such as 'lichen-patched'. Read page 156 again and find other examples of compound words. Then write a description of your own fantastic creature. Describe its size, shape and movements. Include compound words.

Floodland

Before you read

1   Have you ever been on a boat? Would you like to? You may have been on a long voyage. Are you a natural sailor or not? You may even have lived on a houseboat. Discuss your ideas and experiences in a group.

What's it about?

Read the text, then work with a partner to make short notes in answer to questions 2 to 4.

2   What is the importance of the boat for Zoe? What must she get away from? What or who must she find?

3   Zoe says 'she'd made it a rule not to trust anyone' (page 161). Why? What do you think has happened to everyone?

4   Find Norwich on a map of the British Isles. What do you think has happened to Norwich? In which direction is Zoe rowing? Where is she rowing to? Draw a sketch map to show what has happened to Norwich.

Thinking about the text

5   The opening paragraph conjures up a sense of speed, through images such as 'Her feet pounded' (page 159). Find other examples. Then think of your own images of speed using powerful verbs, such as:

   hurtled   propelled   catapulted   slung   sprinted

   Use a thesaurus if you need to. Share your descriptions with a partner.

6   A captain of a ship keeps a ship's log. This is rather like a diary – a record of a journey, detailing everything that happens and other information such as the direction the ship is heading. Write Zoe's log for three days. Describe the events and her feelings. Draw on information from the text and your own imagination.

7   Why do you think Marcus Sedgwick has chosen to write a book called *Floodland*? Working in a small group, think of ten questions to ask him about:
   * the subject of the book
   * where he got his ideas from
   * what he is concerned about.

Two poems

Before you read

1   When homes are flooded the event is often a catastrophe. Have you ever experienced flooding? Discuss with a partner the different effects you think flooding can have on our lives.

What's it about?

Read the text, then work with a partner to make short notes in answer to questions 2 to 4.

2   Where do you think the narrator of *From the Flood Plain* is standing? What does he see around him? What is the extent of the flood?

3   What is the 'gilded bilge' in *From the Flood Plain* and what creatures are at home there? How can you tell the flood isn't over yet?

4   When is *Après* set? What has happened to the flood in the meantime?

Thinking about the text

5   In *From the Flood Plain* there is a battle for territory between humans and nature. Find some words or lines that illustrate this, such as 'will turf us out'. Who do you think is winning?

6   **a**   In *Après*, what are the effects of the flood on the house? The language used often depicts the contents as though they had been in a battle and as though they were human (personification). Jot down some examples of the 'battle scene'. For example, 'the plaster blistered / with salts'.

   b   Write a paragraph explaining how the depiction of the building and its contents illustrates how easily our modern lives are disrupted.

7   **a**   Re-read both poems. Then discuss with a partner nature's capacity to flourish as a result of the flood. For example, we are told that the garden is 'good, and greener' (page 165) – why is this?

   b   Working with your partner, write a poem about nature's ability to renew itself. Organise it as you please. For example, you can write a verse each, or choose the words together. You could think about derelict places and how nature takes over.

Classic Combo

Before you read

1   What is your normal diet? Make a quick list of the kinds of food you eat. Keep your list to refer to later.

What's it about?

2   Why do you think the artist has chosen the title? In what way is it familiar to all of us? Think of some alternative titles to suit the graphic fiction and list them.

3   What do you think is the artist's intention in creating *Classic Combo*? What impact would it have on most readers? Discuss your ideas with a partner.

Thinking about the text

4   Work with a partner. Follow the five stages in the production of *Classic Combo*. As you follow the stages discuss:
   - the resources involved
   - how agriculture is involved
   - how energy is use
   - how transport is involved.

   Which aspect in the production of *Classic Combo* do you find most disturbing and why?

5   Draw up a ten-point plan outlining how you could live in greater harmony with the natural world by changing what you eat and where you buy it. Refer back to the list you made in question 1. Think about these questions.
   - Where does your food comes from? (Check details on packages to find out where it is grown or produced.)
   - Do you buy locally?
   - Does anyone in your family grow food?
   - Do you waste food?

6   Create your own graphic fiction or a design for some graffiti art about one of the following:

   the food we eat   growing food   saving energy

   For example, you could make the case for vegetarianism or humane farming, or focus on problems related to pesticides or the car.

The Urban Farmer

Before you read

1   Have you ever grown anything? How did you look after it? If you have never grown anything, imagine what it must be like. Write a few sentences to describe the satisfaction growing something brings.

What's it about?

Read the text, then work with a partner to make short notes in answer to questions 2 to 4.

2   What is inner-city agriculture? What project has Fritz Haeg headed and for which organisation?

3   How were local residents involved? In what ways was the project successful and why? Identify several reasons, thinking about what was produced and the effect on the local community.

4   Your carbon footprint is the amount of CO_2 you produce through activities such as travel, lighting, heating and so on. What accounts for our greatest use of carbon, according to the World Wildlife Fund? Why does this mean that urban farming is an especially good idea?

Thinking about the text

5   Re-read the text and make notes on how practical it is to turn cities into urban farms, including rearing animals as well as gardening. Note down some examples and decide what might be popular with the public and what might be less popular. Then write your own newspaper report, entitled 'Growing your own'.

6   Imagine you are a gardener in Haeg's project. Present an entertaining talk on your experiences to a small group.

7   How could land in your school be used to grow food? Working in a small group, design a gardening project for your school.
   ● What spaces would you use?
   ● What would you grow?
   ● How could the whole school be involved?
   ● What would you do with the food you grew?

   Appoint someone in the group to make notes. Write up your project, allotting different sections to different members. Use an informative style and include bullet points, diagrams and captions.

The Rough Guide to Climate Change

Before you read
1 If someone says the words 'climate change', what comes to mind? Write down five things connected to climate change.

What's it about?
2 What kind of questions do the three headings (pages 184, 189 and 191) cover? Write a short explanation for each one.

3 Why does the carbon dioxide we emit into the atmosphere have such a great effect? What does the eruption of the volcano Krakatoa show us? Discuss your ideas with a partner and make some notes.

Thinking about the text
4 Design a simple table to display the following. For example, you could start by listing the dates in chronological order.
- The average surface temperature of Earth has risen by 0.7°C over the last century.
- The first six years of the twenty-first century were the hottest.
- Montana's Glacier National Park is expected to lose its glaciers by 2030.
- Arctic sea ice has lost nearly half its average summer thickness since 1950.

5 Work with a partner. Take it in turns to explain these points to each other. Try to take only one or two minutes to explain each point.
- The link between rainforests and CO_2
- How rainforests are destroyed
- Why the destruction of rainforests contributes to climate change
- How rainforests cool the climate locally

Now use your knowledge to create a poster about saving the rainforests.

6 Study the section 'What can we do about it?' (pages 191–3). Identify:
- the difficulties that confront the planet
- possible solutions.

Write an essay based on this information. Use a few quotations from the text to support your ideas.

How Many Light Bulbs Does it Take to Change a Planet?

Before you read

1   Can you remember what the terms 'sustainable development' and 'biodiversity' mean? Check in an encyclopaedia. Discuss the definitions with a partner and between you think of one example of each.

What's it about?

Read the text, then work with a partner to make short notes in answer to questions 2 to 4.

2   What rare experience did the writer have as a member of a research team? How must he have felt? What has brought about 'this desperate state of affairs' (page 194)? Why might there be a happy ending?

3   Why are forests so important to the health of the planet?

4   Re-read the first solution (pages 196–198). How can you tell from the text that Tony Juniper is committed to the environment? Find phrases that reveal this.

Thinking about the text

5   Fens are wetlands found in low-lying areas. How have the fens of East Anglia changed? Draw a timeline recording what has happened from the Roman period to the present day. Write a paragraph to go with your timeline, explaining how and why the Fens are being restored.

6   Refer to the second solution (pages 198–200), which deals with societies that live sustainably. Discuss the questions below in a small group, appointing one person to make notes.
   a   What are transnational corporations? What ecological damage do they do, for instance in Brazil, and why?
   b   If local people retained their rights over their land, how would it benefit the habitats and species? Where is there a good example of this?
   c   What is the Rainforest Foundation?

7   Find out more about The Rainforest Foundation by visiting http://www.rainforestfoundationuk.org. Then, in your group, draw up a charter of rights for forest communities. Present your charter to another group and listen to theirs. Share all your ideas until your class has a final draft.

The Final Straw

Before you read

1   Species become extinct naturally, but at a low rate, as conditions change. Humans have increased this rate dramatically. In what ways do you think we have done this? Discuss your ideas in a small group. Can you think of any examples?

What's it about?

Read the text, then work with a partner to make short notes in answer to questions 2 to 5.

2   The poem presents us with two worlds. One is set in the past. What part of history does it conjure up? Where is the other, 'the new world'?

3   Why are childless couples offered the scythe? What do they hope to release as they cut? What lurks in the 'hub of stalks'? What other things are they offered to help them?

4   What happens when the scythe ('the blade') is used? What emerges from it?

5   The poem *The Final Straw* is the last in Simon Armitage's volume. Why is this a fitting place for the poem?

Thinking about the text

6   **a**   There are potent themes in this poem related to the seasons. Which do these images refer to?
- fat, flowing grain
- We harvested clockwise
- a hub of stalks

   b   The seasons are also related to the life cycle: birth, marriage and death. Find other images of this cycle in the first two verses of the poem.

7   The image of the Spix's macaw disappearing is strong and powerful because we follow its course with our inner eye. But this is not only a bird disappearing from view. What else does the image represent? What part of the life cycle would you link to the Spix's macaw? How might this affect the childless couple? (Look back at your answer to question 3.) Write an appreciation of the last verse, outlining what it means.

The Forgotten Enemy

Before you read

1 We often feel sorry for ourselves or uncomfortable when we are alone. But in what ways can being alone be a beneficial experience? Discuss your ideas with a partner.

What's it about?

2 Where and when is the story set? Find quotations in the text to support your answers.

3 What striking visual images come to mind when you read the story? Which do you find the most strange? Make some notes with a partner.

Thinking about the text

4 What is Professor Millward like? Draw a spider diagram with his name in the centre. List all his characteristics around it. Consider:
   - his main traits and how they have helped him survive emotionally
   - how he has survived physically
   - the things he cares about.

   Write a short character sketch of the Professor. Conclude by assessing his future chances of survival, giving reasons for your assessment.

5 The plot is driven by an unknown 'presence' in the story that helps to build suspense. For example, in the first paragraph (page 205), the reader's curiosity is aroused by the sentence beginning 'This time, he was sure, it had been no dream'. Find other examples throughout the story that do the same. What does the 'presence' turn out to be?

6 'Given what we know about climate change, *The Forgotten Enemy* is no longer relevant today.'
   Carry out a group debate in which some members argue that this statement is true, while others argue that the story is still relevant today.

7 Write a review of *The Forgotten Enemy*. Include sections on:
   - the main character
   - the setting and images
   - why the story captures and retains our attention.

   But remember, don't give away the ending!

Compare and contrast

1 Which texts in this section:
- gave you information you didn't know
- clarified issues you were uncertain of
- affected you emotionally
- stimulated creative ideas?

Discuss these points in a group, modifying your ideas if you need to.

2 Work in a small group to design the home page of a website for children. Choose ideas from the texts in this section, *Future planet*. For example, you could include a picture or drawing of the Spix's Macaw to represent species extinction.
- Write a welcome statement and outline the purpose of the website.
- Create a menu listing different information (for example, What is climate change?).
- Include drawings and diagrams.
- Think of a suitable name for the website.

3 What do *The Final Straw* and the introductory section of *How Many Light Bulbs Does it Take to Change a Planet?* have in common? How do they express their concerns differently? Write a poem, using information from the latter to spark ideas. For example, what do these statements make you think about?
- the single male we had found was the last of his kind (page 194)
- the rarer it got, the more they [the collectors] wanted it (page 194)

4 Draw on the information, issues and themes in this section to write a newspaper article in which you discuss whether we treat the natural world as well as we should. Follow this plan or make your own.
- Write a general introduction about your intentions.
- Point to examples from different texts to make your points over several paragraphs.
- Write a conclusion emphasising the way forward.

Draw on handy connectives to structure your paragraphs; for example:

in addition   as well as   as a result   consequently   however

5 Both *How Many Light Bulbs Does it Take to Change a Planet?* and *The Rough Guide to Climate Change* discuss the importance of rainforests, but they focus on different aspects of them. With a partner, discuss the similarities and differences between the two texts.

Notes on authors

Simon Armitage (1963-) writes in a wide range of genres. He is a poet, dramatist, novelist and lyricist and also writes for television, film and the stage. He has received many awards for his work. His poems are often thoughtful and witty. His most recent publication is a translation of *Sir Gawain and the Green Knight* (Faber and Faber, 2007).

David Attenborough (1926-) was born in Leicester. His interest in the natural world began when, as a child, he collected fossils and other specimens. In 1952 he joined the BBC and his career has spanned over fifty years. He was Director of Programmes for BBC One and Two during the Sixties and Seventies, but returned to making documentary programmes in 1972. The most famous of his programmes are the eight BBC programmes known as *The Life Collection*, on aspects of the natural world.

Sujata Bhatt (1956-) is a well-known Indian poet and translator of Gujurati poetry into English. She was born in Ahmedabad and brought up in India, until her family moved to the USA. She now lives in Bremen, Germany. Her own poems are often concerned with the relationships between different cultures, and the effects of racism. Her poem *Search for My Tongue* is part of the GCSE Anthology section.

Elizabeth Bishop (1911–1979) was an American poet born in Massachusetts and raised by her grandparents. She travelled extensively through Europe and North Africa and lived in Brazil. Her travels are reflected in her poetry, which explores the physical world in a quick-witted, well-crafted style. She was the American Poet Laureate from 1949–1950 and won the Pulitzer prize for poetry in 1956.

Melvin Burgess (1954-) is a children's author who was born in Sussex. He began work as a journalist and did several other jobs before his first book was published. He writes fantasies such as *The Ghost Behind the Wall* (Andersen Press Ltd, 2000) as well as realistic fiction; he also wrote the novel version of Billy Elliot (Scholastic, 2001).

Kate Burt is a freelance journalist. She has written for *The Independent, The Guardian, The Times, Cosmopolitan, Time Out* and *First Magazine*.

John Clare (1793–1864) was a countryman poet, born and bred in Northamptonshire. He began writing poems in early adulthood; he offered them for sale in order to prevent his parents' eviction from their cottage. This eventually led to the publication of his first volume, *Poems Descriptive of Rural Life and Scenery* (Taylor & Hessey, 1820). Clare is still remembered in Northamptonshire, where he is buried in the Churchyard at Helpston, the village of his birth. Children at the local John Clare Primary School lay 'midsummer cushions' – trays of flowers – at his grave on his birthday.

Arthur C. Clarke (1917–2008) was a British author who wrote novels, short stories and non-fiction and was also an inventor. His work was first published in the 1950s. During the 1960s, he and film director Stanly Kubrick wrote the screenplay and novel *2001: A Space Odyssey*. Clarke won the UNESCO–Kalinga prize for the Popularization of Science in 1961 and was knighted in 1998.

Susan Cooper (1935–) has written many books for young people. Her five-volume fantasy novel *The Dark is Rising* (Jonathan Cape) is set in England and Wales and is influenced by British folklore. She was British-born but has lived in America for many years, where she is actively involved in the National Children's Book and Literacy Alliance. Susan Cooper also writes for adults.

Roger Deakin (1943–2006) worked in advertising and teaching and was a writer and filmmaker, travelling widely to other cultures to make films. As an environmentalist, he was a campaigner for the improvement of British rivers. He lived in Suffolk on a farm with a moat, which features in his popular and acclaimed book *Waterlog* (Chatto and Windus, 1999). He completed *Wildwood: A Journey Through Trees* (Hamish Hamilton Ltd, 2007) just before he died.

Berlie Doherty (1943–) is a novelist and playwright and has also written for the screen, although she is best known for her children's books. She was born in Liverpool. She worked as a teacher and social worker before becoming a full-time writer.

Jane Goodall (1934–) was born in London. From an early age she was interested in animals and at only 26 she began to study the social and family life of chimpanzees in Tanzania. In 1977 she established the Jane Goodall Institute, which assists research into chimpanzees. She is also a conservationist concerned with the protection of chimpanzee habitat.

Mike Gould (1960–) is a writer of educational books and books for teenagers. He worked in education for many years and has been a commissioning editor. He now runs Gould Publishing, project managing and writing books and digital resources.

David Heatley (1974–) is an American artist, cartoonist and illustrator. He studied painting and filmmaking and began to draw graphic fiction in his twenties. His work has appeared in *The New Yorker*, *The New York Times* and *McSweeney's*. He is also a musician.

A. L. Hendriks (1922–1992), the Jamaican poet, was a literary critic and a columnist writing for the Jamaican newspaper *The Daily Gleaner*. He also became prominent in Caribbean broadcasting.

Robert Henson's (1960–) *The Rough Guide to Climate Change* (Rough Guides Ltd, 2006) was shortlisted for the 2007 Royal Society Prize for Science Books and is a best seller. He has also written *The Rough Guide to Weather* (Rough Guides Ltd, 2002).

Karen Hesse (1952–) was born in Baltimore, America. She had a variety of jobs for several years, but always kept writing. Many of her books are concerned with social issues and history. Titles include *The Music of Dolphins* (Scholastic, 1996) and *Witness* (Hyperion Books for Children, 2001).

Charlie Higson (1958–) has written and produced television programmes, writing for Harry Enfield and Paul Whitehouse. He is also an actor and comedian. Apart from being the author of the *Young James Bond* novels, he will be writing a new series of books for Puffin.

Barry Hines (1939–) is a British writer, born in Barnsley, who has written several books and scripts for television. He was a teacher before he became a full-time author. His most famous book is *A Kestrel for a Knave* (Michael Joseph, 1968), which was made into a film *Kes* (MGM), directed by Ken Loach, and for which Barry Hines wrote the script. He wrote *The Play of Kes* with Allan Stronach.

Helen Hubbard (1982–) attended Monk's Walk School in Welwyn Garden City when her English teacher, Carol Hedges (who now writes teenage fiction), put forward her poem for the *Times Educational Supplement* competition. Helen works in healthcare management. She still enjoys fiction and poetry and hopes to continue writing in the future.

Ted Hughes (1930–1998) is often named as one of the best poets of his period. He was born in Yorkshire and his early experiences of being close to nature had a profound effect on his writing. He was married to the American poet, Sylvia Plath. Among his collections are *Crow* (Faber and Faber, 1970) and *Birthday Letters* (Faber and Faber, 1998). Ted Hughes also wrote for children: his most famous book is *The Iron Man* (Faber and Faber, 1973). He also compiled an anthology of poetry with the poet Seamus Heaney, called *The Rattle Bag* (Faber and Faber, 1982), which is frequently read in secondary schools.

Tony Juniper (1960–) is an environmentalist, campaigner and ornithologist and is involved in several environmental organisations. He joined Friends of the Earth in 1990 and was the executive director from 2002 to 2008. He has also written *Spix's Macaw: The Race to Save the World's Rarest Bird* (Fourth Estate Ltd, 2002).

Jack London (1876–1916) was an American writer, born in San Francisco. He became very popular in his day. In 1897 he went north to join the Gold Rush in the Klondike. The experience influenced many of his most successful stories. One of Jack London's best-known books is *The Call of the Wild* (William Heinemann, 1903).

Jim Lynch (1961–) grew up in Seattle and after finishing university became a journalist. He wrote for the *Seattle Times* and won awards for his stories. He lives near Puget Sound, the backdrop for his novel *The Highest Tide* (Bloomsbury Publishing Plc, 2005). The novel was adapted into a play for BBC Radio 4 in 2008.

Richard Mabey (1941–) is a naturalist, writer and newspaper columnist. He is the author of many books on the natural world (including *The Concise Flora Britannica,* Chatto and Windus, 1998), some of which have been made into programmes for television. He has produced and narrated his own programmes, such as *Postcards from the Country* (BBC, 1996), and has also written for children. He has won several awards, including the Whitbread prize for his biography of the naturalist Gilbert White (Century Hutchinson, 1986).

Jamie McKendrick (1955–) has written several collections of poetry including *The Sirocco Room* (Oxford University Press, 1991), *The Marble Fly* (Oxford University Press, 1997), *Ink Stone* (Faber, 2003) and, most recently, *Crocodiles and Obelisks* (Faber, 2007). His poetry reviews appear in newspapers and magazines and he is currently writer in residence at the University of London.

Michael Morpurgo (1943–) was born in Hertfordshire. He spent unhappy times in boarding schools, which influenced his writing. As a teacher, his love of storytelling encouraged him to write and he has since won numerous awards. Several of his books have been turned into films and his novel *War Horse* (Scholastic Press, 1982) was adapted for the stage by Nick Stafford. Morpurgo and his wife established the charity Farms for City Children.

Grace Nichols (1950–) was born in Guyana, growing up in the countryside in her early years, before moving to the city. Her poetry was first published as the collection *I is a Long-Memoried Woman* (Carribean Cultural International, 1983). A film adaptation of the book was also made. She has written several volumes of poetry, her most recent being *Startling the Flying Fish* (Virago Press Ltd, 2005). She also writes books for children, influenced by Guyanese folk tales.

Chris Packham (1961–) has been involved with the natural world for many years and has written several best-selling non-fiction books. He was involved with the BBC children's programme *The Really Wild Show* and the nature photography programme *Wild Shots* for Channel 4.

Pliny the Younger (c.61 AD–c.112 AD) was a philosopher, author and lawyer in Ancient Rome. He was born in Como, Italy. His father died when he was a child and he was brought up by his uncle, Pliny the Elder. His letters record his career and life and give us insights into the culture of the period and the changes it underwent.

A. K. Ramanujan (1929–1993) was born in Mysore City to a Tamil-speaking family. Ramanujan himself spoke and wrote in five languages and was an academic writer as well as an important Indian poet. He was Professor of Linguistics at the University of Chicago.

Marcus Sedgwick (1968–) is a British children's writer, an illustrator and also a musician. He was born in Kent. He has published several novels for young people, including *The Dark Horse* (Orion Children's, 2002), which was shortlisted for the Carnegie Medal in 2002. He has also published a picture book, *The Emperor's New Clothes* (Chronicle Books, 2004).

Acknowledgements

The volume editor and publishers acknowledge the following sources of copyright material and are grateful for the permissions granted. While every effort has been made, it has not always been possible to identify the sources of all the material used, or to trace all copyright holders. If any omissions are brought to our notice we will be happy to include the appropriate acknowledgement on reprinting.

p. 2 *My Life with the Chimpanzees* by Jane Goodall, copyright 1988 Byron Preiss Publications, with permission; p. 7 from 'Maninagar Days' by Sujata Bhatt, from *Monkey Shadows*, published by Carcanet Press Limited, 1991; p. 9 *Kite* by Melvin Burgess, Puffin Books, 1997; p. 16 *The Play of Kes* by Barry Hines and Allan Stronach, Heinemann Play Series; p. 26 'The Fish' from *The Complete Poems 1927–1979* by Elizabeth Bishop, copyright © 1979, 1983 by Alice Helen Methfessel, reprinted by permission of Farrar, Straus and Giroux, LLC; p. 29 'The Sudden Knowledge of Moles' by Mike Gould, copyright 2008, reproduced by permission of the author and Gould Publishing Ltd; p. 31 from *Chris Packham's Wild Side of Town* by Chris Packham, published by New Holland; p. 34 excerpt from 'Snakes' from *Striders* by A.K. Ramanujan, 1966, by permission of Oxford University Press; p. 35 *Life in Cold Blood* by David Attenborough, published by BBC Books, 2008; p. 53 from *The White Horse of Zennor* by Michael Morpurgo, text copyright © 1982, published by Egmont UK Ltd London, used with permission; p. 69 from *Daughter of the Sea* by Berlie Doherty, first published by Hamish Hamilton, 1996; p. 74 'The Fringe of the Sea' by A.L. Hendriks; p. 76 'For River' by Helen Hubbard, 1999; p. 78 'Sea Timeless Song' from *The Fat Black Woman's Poems* by Grace Nichols, published by Virago, an imprint of Little, Brown Book Group, copyright © Grace Nichols 1984, reproduced with permission of Curtis Brown Group Ltd; p. 79 from *The Highest Tide* by Jim Lynch, published by Bloomsbury; p. 84 from *Waterlog - A Swimmer's Journey through Britain* by Roger Deakin, published by Chatto & Windus, reprinted by permission of The Random House Group Ltd; p. 100 'Dust Storm' by Karen Hesse from *Out of the Dust* published by Francis Lincoln Ltd, copyright © 2007, produced by permission of Frances Lincoln Ltd; p. 105 *Hurricane Gold* by Charlie Higson, copyright © Ian Fleming Publications Ltd 2007, reprinted with permission from Ian Fleming Publications Ltd; p. 127 'Wind' by Ted

Hughes from *The Hawk in the Rain*, published by Faber and Faber Ltd; p. 131 Letter 6.16 from Pliny the Younger to Tacitus, translated by Professor Cynthia Damon; p. 136 from *Nature Cure* by Richard Mabey, published by Chatto & Windus, reprinted by permission of The Random House Group Ltd; p. 151 from *Green Boy* by Susan Cooper, published by Bodley Head, reprinted by permission of The Random House Group Ltd; p. 159 from *Floodlands* by Marcus Sedgwick, published by Orion Children's Books, a division of The Orion Publishing Group; p. 164 'From the Flood Plain' and 'Après' by Jamie McKendrick © Jamie McKendrick; p. 166 'Classic Combo #3' by David Heatley, published in *Granta 102*, Granta Publications; p. 177 'The urban farmer: One man's crusade to plough up the inner city' by Kate Burt, published in *The Independent*, 1 June 2008; p. 184 from *The Rough Guide to Climate Change* by Robert Henson (Rough Guides 2006), copyright © Robert Henson, 2006, reproduced by permission of Penguin Ltd; p. 194 *How Many Lightbulbs Does it Take to Change a Planet?* By Tony Juniper, published by Quercus, 2007; p. 204 'The Final Straw' by Simon Armitage, from *Tyrannosaurus Rex versus the Corduroy Kid*, published by Faber and Faber Ltd, copyright © Simon Armitage, 2003; p. 205 from *Collected Stories* by Arthur C Clarke, published by Gollancz.

The publishers would like to thank the following for permission to reproduce photographs: p. 23 Moviestore Collection.com; p. 37 © Chris Mattison, Frank Lane Picture Agency / Corbis; p. 75 © Donald Nausbaum / Corbis; p. 76 Jupiter Images; p. 82 Christian Darkin / Science Photo Library; p. 84 Andrey Armyagov / Shutterstock; p. 101 Hulton Archive / Getty Images; p. 113 Theo Allofs / Photolibrary; p. 132 Mt Vesuvius erupting in March 1944, photo by John Reinhardt, B24 tailgunner in the U.S. Air Force; p. 185 NASA; p. 195 © Nick Gordon / Naturepl.